INTRODUCTORY CHEMISTRY FOR THE ENVIRONMENTAL SCIENCES

CAMBRIDGE ENVIRONMENTAL CHEMISTRY SERIES

Series Editors:

P.G.C. Campbell, *Centre for Advanced Analytical Chemistry, CSIRO, Australia*

J.N. Galloway, *Department of Environmental Sciences, University of Virginia, USA*

R.M. Harrison, *Department of Chemistry, University of Essex, England*

Other books in this series:

1 P. Brimblecombe *Air Composition & Chemistry*

2 M. Cresser and A. Edwards *Acidification of Freshwaters*

3 A.C. Chamberlain *Radioactive Aerosols*

This undergraduate textbook has been written for students specialising in environmental science or ecology and who require an introduction to the basic chemical concepts that underlie a sound understanding of the subject. The aim throughout is to provide an introduction to basic chemical principles, starting with atomic and molecular structure and leading to those more advanced areas within physical, organic, inorganic and analytical chemistry which are fundamental to a full understanding of environmental chemistry. Wherever possible, these principles are illustrated with examples from environmental science. The final chapter provides case studies of environmental chemical processes. Whilst the primary aim is to satisfy the needs of the non-chemist, it will also be useful to chemists who wish to apply their expertise to new topics within environmental science and ecology.

The volume is suitable for undergraduates at universities, colleges and polytechnics studying courses in environmental science, environmental chemistry and ecology.

INTRODUCTORY CHEMISTRY
FOR THE ENVIRONMENTAL
SCIENCES

ROY M. HARRISON
Department of Chemistry, University of Essex

S.J. de MORA
Department of Chemistry, University of Auckland

S. RAPSOMANIKIS
Department of Biogeochemistry, Max-Planck-Institute, Mainz

W.R. JOHNSTON
Fiber-Seal (UK) Ltd, Bradford

CAMBRIDGE
UNIVERSITY PRESS

Published by the Press Syndicate of the University of Cambridge
The Pitt Building, Trumpington Street, Cambridge CB2 1RP
40 West 20th Street, New York, NY 1011-4211, USA
10 Stamford Road, Oakleigh, Melbourne 3166, Australia

First published 1991
Reprinted 1993

Printed in Great Britain at the University Press, Cambridge

British Library cataloguing in publication data

Introductory chemistry for the environmental sciences.
1. Chemistry
I. Harrison, R.M.   II. Series
540

Library of Congress cataloguing in publication data

Introductory chemistry for the environmental sciences/R.M. Harrison
[et al.].
    p.   cm.—(Cambridge environmental chemistry series)
Includes bibliographical references.
ISBN 0 521 25673 9. — ISBN 0 521 27639 X (paperback)
1. Environmental chemistry.   I. Harrison, R.M.   II. Series.
QD31.2.I572   1991
540—dc20   90-2370   CIP

ISBN 0 521 25673 9 hardback
ISBN 0 521 27639 X paperback

PN

CONTENTS

○ ○ ○ ○ ○ ○ ○ ○ ○ ○ ○ ○ ○ ○ ○ ○ ○ ○ ○ ○

PREFACE

All four authors of this book are involved in teaching and/or research in environmental science and ecology. In the course of this work, we have found no shortage of advanced books in specialised aspects of our subject, and the availability of teaching texts on specific aspects of environmental chemistry has improved markedly in recent years. There does, however, in our view continue to be a need for a basic book covering the chemistry necessary to comprehend more specialised books on chemical aspects of environmental science and ecology. This book is intended to fulfil that need.

In preparing the book, we have been struck by the enormous range of aspects of chemistry which are involved in studying the environment; all major branches of the subject are involved to some degree. This has made our task more difficult and explains the involvement of such a large band of authors, needed to give expert coverage to all the topics included. Although specific individuals have prepared first drafts of whole or part chapters, these have been reviewed by all other authors, and the finished work is a group effort.

The book is not aimed at specialist chemists, although we hope that they may find some of the more applied material useful. Rather, it is aimed at students specialising in environmental science or ecology, and requiring a grounding in basic chemical concepts to make chemical aspects of their studies accessible. At all times, except in the introductory Chapter 1, the relevance of each concept to the environment is explained to give added relevance for the student. The final chapter presents case studies drawn from environmental science as an illustration of the concepts covered earlier in the book.

Roy M. Harrison, Colchester

1

○ ○ ○ ○ ○ ○ ○ ○ ○ ○ ○ ○ ○ ○ ○ ○ ○ ○ ○ ○

Introduction

(A)   Atomic and molecular structure

1.1   The atom

For many years it was speculated that the ultimate state of subdivision of matter was discrete indivisible particles termed atoms. In the early nineteenth century John Dalton gave quantitative support to these ideas by assigning a self-consistent set of masses to atoms of elements. Elements are substances which cannot be decomposed by ordinary types of chemical change or made by chemical union. However, subsequent work by Faraday, Crookes and Goldstein pointed out the electrical nature of atoms and that they are divisible. Our simple picture of an atom now shows a central nucleus surrounded by a cloud of electrons.

1.1.1   Electrons

When a neutral atom is supplied with sufficient energy it will ionise yielding positively and negatively charged fragments. The former, the positive ion, is characteristic of the particular parent atom, whilst the negative fragment, the electron, is the same irrespective of the parent atom. The charge and mass of electrons can be determined by studying their behaviour in electrical and magnetic fields. Such experiments were first performed by J.J. Thompson in 1897. Electrons produced at a cathode were accelerated towards a positive anode, in a vacuum tube, and focused into a narrow beam, which struck a zinc sulphide screen, which glowed, indicating the beam position. The beam could be deflected in a vertical direction by varying the voltage between two plates above and below the electron beam. A magnet surrounding the evacuated tube could bend the electron beam into a curved path. J.J. Thompson's experiments consisted of first measuring the electron beam deflection due to the magnetic field alone and then applying a potential across the plates to restore the beam to its original path; i.e. the magnetic and electric fields counteract each other.

An electron, with charge e, moves with velocity v through a magnetic field of flux density B, and is deflected into a curved path with radius r (Figure 1.1). The force of the magnetic field is given by Bev and is equal to the product of the electron mass, m, and its acceleration, v^2/r:

$$Bev = mv^2/r \qquad (1.1)$$
$$e/m = v/Br \qquad (1.2)$$

Since B and r can be measured, knowing v would enable the charge to mass ratio to be determined. To find v, the electric field strength, E, which restores the electron beam to its original path, is related to the magnetic force by

$$Bev = Ee \qquad (1.3)$$

(magnetic force = electric force)

therefore

$$v = E/B \qquad (1.4)$$

and by substitution into equation (1.2)

$$e/m = E/B^2r \qquad (1.5)$$

The value of e/m for all electrons is 1.7588×10^8 coulombs per gram $(C\,g^{-1})$.

To evaluate the mass of an electron, experiments were carried out to determine the applied potential required to prevent oil droplets

Figure 1.1. Path of electron moving through a magnetic field.

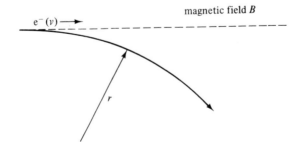

magnetic field B

$e^-\ (v)$

carrying negative charges from settling due to gravity. This enabled the charge on the droplets to be determined, and these were found to be whole number multiples of 1.6×10^{-19} C. It was assumed that the smallest charge held by an oil drop was 1.6×10^{-19} C which represented the charge of a single electron. Combining the e/m value with the charge of an electron gives the mass of an electron to be 9.1×10^{-28} g.

1.1.2 Nucleus

Early ideas based on Thompson's findings suggested that atoms consisted of electrons and positive components equally dispersed in space. Experiments performed by Lord Rutherford in 1911 cast doubt on such a model of matter. He projected alpha particles (helium atoms from which two electrons have been removed) from a radioactive source towards a metal foil sheet and measured their deflection. If the positive and negative charges were distributed evenly throughout the metal foil, then alpha particles would essentially pass undisturbed through the target. However, Rutherford found that many alpha particles emerging from the metal foil had been considerably deflected. Rutherford accounted for his observations by suggesting that positive charge and mass are highly concentrated in discrete areas of the foil (and in all matter). When an alpha particle encounters such a discrete unit it is deflected. It was therefore postulated that atoms have a positively charged nucleus or discrete centre surrounded by the negative electrons. We now know that the positive charge of the nucleus is due to the presence of protons, each of which has a charge equal but opposite to that of the electron.

1.1.3 Atomic number

Atomic number, usually designated Z, represents the magnitude of positive charge in the nucleus and is equal to the number of positively charged particles, known as protons. Elements appear to have ordered chemical behaviour according to their atomic number (Section 1.1.6). The pioneering work of H.G.T. Moseley established that the square root of X-ray frequency emitted from elements bombarded with electrons gave a linear relationship when plotted against the order in which elements appeared in the periodic table (see Section 1.1.6). The interaction of electrons with the positive charge in each element's nucleus determines the characteristic X-ray frequencies. Moseley left gaps in his ordering of elements, in terms of atomic number, which corresponded to, at that time, elements still to be discovered.

1.1.4 Atomic mass

Chemical behaviour closely relates to atomic number, i.e. the total number of positive charges associated with each element's nucleus, whereas changes in atomic mass (i.e. the total mass of the atom) often are not related to chemical behaviour. This is because different atoms of the same element can have different masses and are said to exist as isotopes.

Most of the mass of the atom is associated with the nucleus. Thus changes in mass between atoms of the same element cannot be attributed to differences in mass associated with positive charge alone. Moving from carbon of mass 12 (^{12}C) to carbon of mass 13 (^{13}C) results in a mass change of one atomic mass unit, whilst the atomic number remains unchanged at six. We account for this by postulating that the nucleus is composed of two main components – protons, which impart mass and positive charge, and neutrons, which are uncharged but have a mass approximately equal to that of a proton.

Thus isotopes of the same element have identical numbers of protons, i.e. the same atomic number, and differ from each other in the number of neutrons in their nuclei. For example, nitrogen-12 (^{12}N) has seven protons (atomic number 7) and five neutrons (atomic mass = $7 + 5 = 12$), whilst ^{13}N has seven protons and six neutrons. Nuclei are usually represented in shorthand notation as follows:

$$^{12}_{7}\mathrm{N}$$

The subscript 7 denotes the atomic number (Z) corresponding to the number of protons in the atom, also the number of electrons in a neutral atom. The superscript 12 indicates the mass number (A). This is the integer sum of protons + neutrons and is approximately the atomic mass of the isotope.

1.1.5 Electrons in atoms
Bohr atom

In 1913 Niels Bohr developed the first theory explaining atomic structure. This sought to explain why negative electrons moving about a positive nucleus do not continuously radiate or absorb energy and do not collapse into the nucleus. Bohr equated the force of electrical attraction between an electron and the nucleus to the centripetal force of the electron. He restricted the angular momentum of electrons in all atoms to specific integer values referred to as the quantum condition.

This can be explained as follows:

$$mvr = nh/2\pi \tag{1.6}$$

where

mvr = electron angular momentum,
n = integer value for quantum condition,
h = Planck's constant.

The integer values (n), referred to as the principal quantum number, have values of $1, 2, 3 \ldots$, etc., where 1 would represent the lowest energy state (ground state) of the atom and higher integers correspond to excited energy states.

Bohr's theory describes the hydrogen atom well, explaining the atomic spectrum (*vide infra*) of this element when elevated to higher energy states. It does not work so well for other elements when electron–electron interactions occur and more sophisticated theories have been advanced, as will be discussed shortly.

Extensions to the Bohr model

Evidence for the arrangement of electrons in atoms has been gained from two major sources.

(a) Ionisation potentials – the amount of energy required to remove an electron from the atom.

(b) The light absorption and emission spectra of elements.

Electrons in atoms can be regarded as delocalised clouds of negative electricity in which there is a statistical probability of finding the electron in any specific space around the nucleus. When driven out of atoms by supplying them with energy, electrons behave both as particles and as waves, similar to light waves but of a much shorter wavelength. This concept was postulated by Louis de Broglie in 1924 in order to explain the quantised nature of electrons in atoms put forward by Bohr.

The wavelength of the de Broglie wave is given by

$$\lambda = h/mv \tag{1.7}$$

where

h = Planck's constant,
mv = momentum of the electron.

de Broglie's postulation was that a stable electron orbit around a nucleus incorporated an integral multiple of wavelengths along the circumference of the orbit. This can be represented as

$$n\lambda = 2\pi r \tag{1.8}$$

where

n = integer,
λ = wavelength,
r = radius of orbit.

We can substitute $\lambda = h/mv$ from equation (1.7) and get

$$n(h/mv) = 2\pi r \tag{1.9}$$

which can be rearranged to give

$$mvr = nh/2\pi \tag{1.10}$$

which is the quantum condition of Bohr given in equation (1.6).

Heisenberg in 1927 presented his famous 'uncertainty principle' relating to particles of low mass, such as electrons. He stated that the uncertainty in the position of an electron around a nucleus multiplied by the uncertainty in the momentum of the electron is of the order of Planck's constant. Accordingly, if we precisely specify momentum then the more uncertain we are of the electron's position about the nucleus. Essentially, this has led to the concept of electrons residing in probability clouds around the nucleus, i.e. electrons do not orbit the nucleus in definite tracks but should be thought of as occupying an area of space in which there is a high probability of locating them. Calculation of such probabilities is the concern of wave mechanics and is mathematically very complex.

Each electron in an atom possesses a certain amount of energy; however, there are limitations on the specific amount. An energy continuum does not exist but rather a small number of discrete energy levels are permitted. That the quantities of energy are definite and cannot have intermediate values is deduced from absorption spectra of elements. When irradiated with light over a continuous spectrum, atoms of a single element absorb only certain wavelengths associated with the energy levels of their electrons. This is explained by ground state electrons taking up discrete quantities of light energy and being elevated to an excited state. The discrete amounts of energy are called quanta. Electrons therefore move from areas of space around the nucleus associated with low energy to areas, or probability clouds, associated with higher energy. It is customary to use the term 'orbital' when referring to an energy level associated with a given electronic probability distribution.

Quantum numbers

From wave mechanics a complete description of each electron in an atom has emerged in terms of four descriptive numbers termed quantum numbers. The Pauli exclusion principle states that no two electrons in the same atom can have the same four quantum numbers.

Principal quantum number (n)
The basic principle of quantum theory is that only specified energy levels are available to electrons in atoms. The energy levels are numbered 1 (if lowest energy), 2 (next higher), 3, etc. The number (n) assigned to a specific energy level in the atom of a particular element is the principal quantum number. The maximum number of electrons in any particular energy level in an atom is given by $2n^2$. Electrons in the lowest energy level $(n = 1)$ require the greatest amount of energy to be removed from the atom.

The electron population associated with each principal quantum number is often said to reside in a particular 'shell'. Shells are usually designated with letters K, L, M, etc., representing principal quantum numbers 1, 2, 3, etc. The letter designation is an inheritance from early work carried out by X-ray spectroscopists.

Angular momentum quantum number (l)
The angular momentum quantum number (l) is a measure of the orbital angular momentum of an electron, determining the shape of its orbital. Each shell is divided into subshells designated as s, p, d and f. They arise from the fact that for each shell the angular momentum quantum number can take values from zero to $n - 1$, and $l = 0$ designates the s subshell; $l = 1$ the p subshell; $l = 2$ the d subshell; and $l = 3$ the f subshell.

For subshells with $l > 0$ (i.e. except s orbital), there are $2l + 1$ equivalent ways that the electron cloud can orientate itself in space. These orbitals then are equivalent in energy and are termed *degenerate*. Taking as an example the p orbitals, they can have three spatial orientations (remember $2l + 1 = 3$), i.e. along the x axis (p_x), along the y axis (p_y), and along the z axis (p_z). Hence the maximum electron density can take three equivalent orientations along any of the three axes (see Figure 1.4). Upon application of a magnetic or electric field, however, the three orientations are not equivalent and the degeneracy of the three p orbitals no longer exists (see also Section 1.2.5).

The arrangement of energy levels described by the first two quantum numbers is illustrated in Figure 1.2.

Magnetic quantum number (m_l)
Orbitals within the same subshell have the same energy but are orientated about the nucleus in different ways. The magnetic quantum number describes the particular orbital within a given subshell that an electron occupies. The magnetic quantum number is so called because a magnetic field will resolve orbitals of the same subshell into different energy levels. Assignments of m_l are the integers from $+l$ to $-l$ (including zero). For example, for $l = 1$ (p subshell), m_l takes values of -1, 0 and $+1$, i.e. there are three p orbitals designated p_x, p_y and p_z.

Within each subshell the shape of electron distribution clouds differs. A full s subshell contains two electrons, the distribution of which is spherical (Figure 1.3). A full p subshell contains six electrons, grouped in pairs occupying particular spatial distributions or orbitals as shown in Figure 1.4. The overall electron distribution of a full p subshell is that of a sphere. The d subshells, when full, contain five pairs of electrons occupying five spatial configurations (Figure 1.5), the sum of which is again spherical.

Figure 1.2. Arrangement of energy levels described by the first two quantum numbers. Values of *l* are in parentheses.

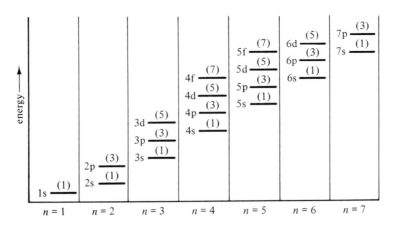

Spin quantum number (m_s)

Since the Pauli exclusion principle states that no two electrons in the same atom can be identical, then each pair of electrons with the same values of n, l and m_l in a full orbital must be distinguished. This is achieved on the basis of the direction of spin about the electron's own axis. Each electron in the orbital is assigned a spin of $+\frac{1}{2}$ or $-\frac{1}{2}$, representing the two possible spin directions. Spins are usually indicated in shorthand by ↑ or ↓. Two electrons with identical values of quantum numbers n, l and m_l are of identical energy (i.e. degenerate) unless an external magnetic or electric field is applied, splitting the energies of the two spins which are otherwise equivalent.

Filling of subshells

The filling of subshells occurs in a logical manner although perhaps not the simplest. Shells are not filled one at a time but instead follow the so-called 'Aufbau principle', which in effect means that

Figure 1.3. Spherical distribution of electrons in a full s subshell.

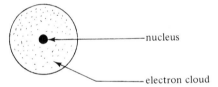

Figure 1.4. Spatial distributions of electron pairs in three p subshells.

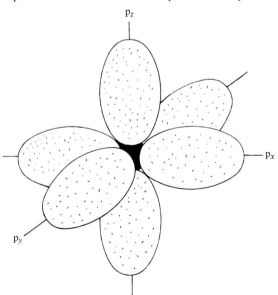

Figure 1.5. Five spatial arrangements of electron pairs in a d orbital.

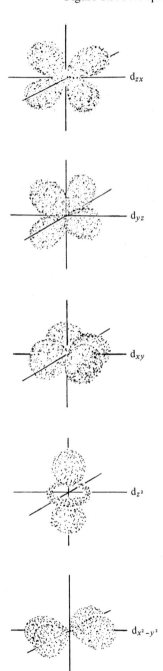

d_{zx}

d_{yz}

d_{xy}

d_{z^2}

$d_{x^2-y^2}$

electrons enter the lowest energy level available. This is illustrated in Figure 1.2 where the 4s subshell is filled before the 3d subshell. A further representation of subshell filling is given in Fig. 1.6, which includes configurations up to $n = 7$.

An important point to remember is that Figure 1.6 is an 'order of filling diagram' and not an energy diagram since no single energy diagram applies to all elements. Figure 1.6 represents the relative order of subshells at the point when filling of orbitals starts to occur. For example, Figure 1.2 indicates 4s lies below 3d. The 4s subshell starts to fill at potassium and the 3d at scandium, and for these elements Figure 1.2 holds true. However, with elements of higher nuclear charge, for example zinc, additional electron interactions result in 3d being lower in energy than 4s. In filling such orbitals, generally one electron is put into separate orbitals of equal energy before pairing occurs. In carbon atoms ($Z = 6$), the 1s and 2s orbitals each have a pair of electrons. The final two electrons are distributed individually in the 2p orbitals (e.g. p_x and p_y) rather than as a pair in any one of the 2p orbitals.

An example of electron configuration is given in Fig. 1.7 and follows the convention shown below.

Figure 1.6. Representation of subshell filling.

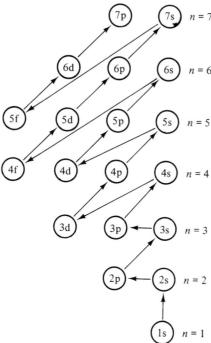

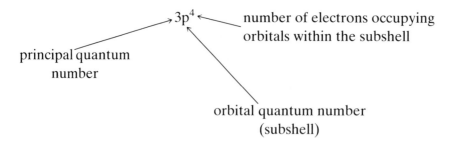

principal quantum
number

$3p^4$ number of electrons occupying
orbitals within the subshell

orbital quantum number
(subshell)

The inert, or noble, gases, neon, argon, krypton, xenon and radon, all have full s and p subshells in their L, M, N, O, and P shells, respectively, and are particularly unreactive; helium has a full K shell.

Figure 1.7. Electronic configuration of various elements. (∗) Filling criteria now start to be influenced by the fact that half filled or fully filled subshells are particularly stable.

H	$1s$						
He	$1s^2$						
Li	$1s^2$	$2s$					
Be	$1s^2$	$2s^2$					
B	$1s^2$	$2s^2$	$2p$				
Ne	$1s^2$	$2s^2$	$2p^6$				
Ar	$1s^2$	$2s^2$	$2p^6$	$3s^2$	$3p^6$		
K	$1s^2$	$2s^2$	$2p^6$	$3s^2$	$3p^6$	$4s$	
Ca	$1s^2$	$2s^2$	$2p^6$	$3s^2$	$3p^6$	$4s^2$	
∗Sc	$1s^2$	$2s^2$	$2p^6$	$3s^2$	$3p^6$	$3d$	$4s^2$
Ti	$1s^2$	$2s^2$	$2p^6$	$3s^2$	$3p^6$	$3d^2$	$4s^2$
V	$1s^2$	$2s^2$	$2p^6$	$3s^2$	$3p^6$	$3d^3$	$4s^2$
Cr	$1s^2$	$2s^2$	$2p^6$	$3s^2$	$3p^6$	$3d^5$	$4s^1$
Mn	$1s^2$	$2s^2$	$2p^6$	$3s^2$	$3p^6$	$3d^5$	$4s^2$
Ni	$1s^2$	$2s^2$	$2p^6$	$3s^2$	$3p^6$	$3d^8$	$4s^2$
Cu	$1s^2$	$2s^2$	$2p^6$	$3s^2$	$3p^6$	$3d^{10}$	$4s^1$
Zn	$1s^2$	$2s^2$	$2p^6$	$3s^2$	$3p^6$	$3d^{10}$	$4s^2$

We have considered the allocation of electrons to subshells in terms of electronic configurations such as, for example, sodium $1s^2 2s^2 2p^6 3s^1$. An alternative representation often used is to show electrons in outermost shells as dots or some other symbol arranged about a central nucleus. Electrons in inner shells are ignored. Examples are given in Figure 1.8.

Magnetism and electron spin

Some materials are weakly attracted to magnets; for example, copper sulphate and oxygen gas. These materials are termed paramagnetic. Other materials are weakly repelled by magnets, for example sodium chloride, and are termed diamagnetic. Materials strongly attracted to magnets are called ferromagnetic, for example iron. All these properties arise from the presence of electrons in atoms.

Electrons in atoms spin about their axes in a clockwise or anticlockwise direction. Paired electrons in any one orbital will have opposing spin directions (Pauli exclusion principle). Any spinning charge creates a magnetic field in its vicinity. If electrons are paired together the opposing spin direction of each electron gives a resultant zero magnetic field. If the electron is unpaired then its magnetic field will prevail, imparting magnetic properties to the atom.

An element, therefore, which has an unpaired electron will be

Figure 1.8. Representation of electrons in outermost subshells of various elements.

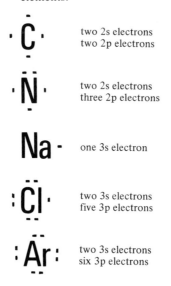

two 2s electrons
two 2p electrons

two 2s electrons
three 2p electrons

one 3s electron

two 3s electrons
five 3p electrons

two 3s electrons
six 3p electrons

paramagnetic. This usually implies an odd number of electrons. However, atoms with even numbers of electrons can also be paramagnetic provided there is an unfilled subshell of electrons allowing parallel unpaired spins. When all electrons in an atom are paired there is no paramagnetism, only diamagnetism, which occurs in all matter due to the actual charge on each electron.

1.1.6 Periodic law and the periodic table

The periodic law was discovered by the Russian scientist Mendeleev some time between 1868 and 1870. In its modern conception, it states that properties of elements recur periodically when elements are arranged in order of increasing atomic number. Mendeleev first devised periodic tables which are essentially similar to those used today, as shown in Table 1.1. Vertical sequences of elements in the periodic table are called groups, whilst horizontal sequences are termed periods. The reason, given by Mendeleev, for the ordering of elements within the table was the similarity of valencies (see Section 1.2.2) of member elements of a group. We can now account for these similarities in terms of electronic configurations of atoms.

In rationalising the periodic table in terms of chemical reactivity of elements in particular groups, a convenient place to start is with group 0 elements which are the so-called inert, noble or rare gases. These are particularly unreactive elements (reactions are known only with fluorine and oxygen) due to the high stability imparted by having full s and p subshells in their outermost shells.

Group IA elements which directly follow the noble gases all exhibit 'metallic' properties, i.e. they have a shiny lustre and are excellent conductors of heat and electricity. These elements have a single valence electron (occupying an s subshell) outside a noble gas configuration. Loss of this single s electron gives a stable electronic configuration to the resultant singly charged positive ion (valency 1), and it is this which essentially governs the chemistry of this group of elements. Group IA elements all react vigorously with water, liberating hydrogen and forming basic solutions which can be neutralized with hydrochloric acid giving a salt.

Group VIIB elements precede the noble gases and exhibit properties characteristic of 'non-metals', i.e. they are poor conductors of heat and electricity. They are collectively called halogens and all react with hydrogen, forming compounds which dissolve in water giving acid solutions (hydrogen fluoride and hydrogen chloride are strong acids).

Table 1.1. *Periodic table of the elements*

Period	Group Ia	Group IIa	Group IIIa	Group IVa	Group Va	Group VIa	Group VIIa	Group VIII			Group Ib	Group IIb	Group IIIb	Group IVb	Group Vb	Group VIb	Group VIIb	Group 0
1 1s	1 H																1 H	2 He
2 2s2p	3 Li	4 Be											5 B	6 C	7 N	8 O	9 F	10 Ne
3 3s3p	11 Na	12 Mg											13 Al	14 Si	15 P	16 S	17 Cl	18 Ar
4 4s3d 4p	19 K	20 Ca	21 Sc	22 Ti	23 V	24 Cr	25 Mn	26 Fe	27 Co	28 Ni	29 Cu	30 Zn	31 Ga	32 Ge	33 As	34 Se	35 Br	36 Kr
5 5s4d 5p	37 Rb	38 Sr	39 Y	40 Zr	41 Nb	42 Mo	43 Tc	44 Ru	45 Rh	46 Pd	47 Ag	48 Cd	49 In	50 Sn	51 Sb	52 Te	53 I	54 Xe
6 6s (4f) 5d 6p	55 Cs	56 Ba	57* La	72 Hf	73 Ta	74 W	75 Re	76 Os	77 Ir	78 Pt	79 Au	80 Hg	81 Tl	82 Pb	83 Bi	84 Po	85 At	86 Rn
7 7s (5f) 6d	87 Fr	88 Ra	89** Ac															

*Lanthanide series 4f	58 Ce	59 Pr	60 Nd	61 Pm	62 Sm	63 Eu	64 Gd	65 Tb	66 Dy	67 Ho	68 Er	69 Tm	70 Yb	71 Lu
**Actinide series 5f	90 Th	91 Pa	92 U	93 Np	94 Pu	95 Am	96 Cm	97 Bk	98 Cf	99 Es	100 Fm	101 Md	102 No	103 Lr

Their chemistry is governed through the acquisition of a single electron giving a noble gas configuration to the outermost shell and a uni-negative charge.

Elements in groups between IA and VIIB show progressive gradation of properties between the two extreme 'metallic' and 'non-metallic' groups.

1.1.7   Atomic radii

It has been pointed out that electrons do not travel at a fixed distance from the nucleus. They can be found almost anywhere between the nucleus and, in theory, infinity. However, the most probable location where an electron may be found can usually be calculated and is relatively near the nucleus. It is almost impossible to define the exact extent of an isolated atom, i.e. to assign a radius to the atom. In addition the electron probability distribution is affected by the environment in which an atom exists, and the size of the atom may change depending on circumstances, for example the nature of other atoms with which it is in combination or close proximity. Atoms in the solid state occupy equilibrium positions in space in which forces of attraction (chemical bonds) balance repulsive forces (from positively charged nuclei). The atoms can then be considered to have finite size imposed by neighbouring atoms. The concept of an atomic radius has some meaning under such circumstances, and if radii can be assigned to atoms of different elements then inter-atomic distances, i.e. bond lengths, of compounds can be predicted. Inter-atomic spacing (the distance between centres of adjacent atoms) can be determined from X-ray diffraction and spectral studies of bound atoms.

With an increase in atomic number (and therefore the number of filled shells around the nucleus), the atomic radii of elements increase moving down any vertical group of the periodic table. This is because the outermost electrons of successive elements in each vertical group occupy shells of increasing principal quantum number. In horizontal periods, the principal quantum number of successive elements does not increase, and in fact there is a gradual decrease in atomic radii moving from left to right along a period as atomic number increases.

1.1.8   Ionisation potential

The energy required to remove completely an electron from the attractive force of the nucleus of a neutral atom is termed the ionisation potential. It is usually measured in electron volts (one electron volt, eV, is the energy gained by one electron accelerated through a potential difference of one volt: $1 \text{ eV} = 1.6 \times 10^{-19}$ J), or may be expressed in kJ mol^{-1}. When considering bonding in subsequent sections, the ionisation potential of the outermost electrons is very important. The ionisation potential of a particular element is determined by factors such as nuclear charge, distance of outer electrons from the nucleus and the screening effect of electrons in lower energy shells.

The higher the nuclear charge, the more difficult is the removal of an electron from the atom, resulting in a high value for the ionisation potential. The further an electron is from the positively charged nucleus, the weaker the force holding them together and therefore the lower the ionisation potential.

Within the periodic table, with increasing atomic number moving down a vertical group, both distance of valence electrons (those principally involved in bonding) from the nucleus and nuclear charge increase. These trends have opposing effects on ionisation potential. Generally, distance effects appear more important since usually ionisation potential decreases with increase in atomic number down a vertical group. On moving across a horizontal period in the periodic table, atomic radius decreases and nuclear charge increases. Both factors increase the first ionisation potential (energy required to remove the outermost electron) of elements within a horizontal group.

Second and third ionisation potentials of elements rise dramatically, since having removed the first or outermost electron, the nuclear charge becomes effectively greater since the electron screen decreases and the remaining electrons are pulled closer to the nucleus.

Ionisation potentials of some transition metals are higher than expected. These elements contain d electrons and it is thought that these are less efficient than s and p subshell electrons at screening valency electrons from the nucleus.

1.1.9 Electron affinity

This is defined as the energy released when a negative ion is formed by addition of an electron to a gaseous atom. Electron affinity measures the tightness of binding of an additional electron to an atom. The halogens (group VIIB elements) have high electron affinities since addition of a single electron to the neutral atom results in formation of an outer shell containing full s and p subshells. This configuration is particularly stable (known as a stable octet). The halogens appear exceptional in their high electron affinity since for most other elements an input of energy is required to add an electron to a neutral atom.

1.1.10 Electronegativity

This is a measure of the tendency of an atom to attract an additional electron towards itself. Numerical values of electronegativity of elements are given in Table 1.2. Electronegativity is an important quantity in determining the nature of bonds between elements and will be considered further in following sections on chemical bonding.

Moving from left to right across a period in the periodic table, effective nuclear charge increases and so does electronegativity. Fluorine, for example, has the highest electronegativity assigned to it of any element in the periodic table. Elements to the left of any period have lower electronegativity. Generally, electronegativity decreases moving down a group since the atomic radius increases.

1.2   Chemical bonds

Two or more atoms held together strongly enough to be considered a single unit are called a molecule, and the attraction between component atoms is called a chemical bond.

1.2.1 Electrons in molecules

Each electron in an isolated atom is under the influence of the nuclear charge and the charges of all of the other electrons present. Within atoms coming together forming molecules, a rearrangement of electrons occurs such that electrons of one atom are influenced by the nucleus of the other atom (i.e. electrons are shared or transferred). If this rearrangement produces an energetically stable condition then bond formation can occur. There are two main theories which attempt to explain the arrangement of electrons in molecules – valence bond theory (VBT) and molecular orbital theory (MOT).

VBT was developed by Linus Pauling. The theory considers atoms in molecules to behave like isolated atoms except that one or more electrons – the valence electrons – from the outer shell of one atom reside in the outer shell of the second atom. Thus a simple diatomic molecule is visualised as two isolated atoms, in close proximity to each other, in which valence electrons of one atom spend part of the time in the outermost shell of the second atom. This theory can account well for the structure and magnetic properties of metal complexes. VBT is considered by many chemists to be of great value since it presents a

Table 1.2. *Pauling electronegativity coefficients χ, as revised by Allred (1961). The roman numeral at the top of each group is the oxidation state to which the coefficients apply*

I	II	III	II	II	II	II	II	II	II	I	II	III	IV	III	II	I
H 2.20																
Li 0.98	Be 1.57											B 2.04	C 2.55	N 3.04	O 3.44	F 3.98
Na 0.93	Mg 1.31											Al 1.61	Si 1.90	P 2.19	S 2.58	Cl 3.16
K 0.82	Ca 1.00	Sc 1.36	Ti 1.54	V 1.63	Cr 1.66	Mn 1.55	Fe 1.83	Co 1.88	Ni 1.91	Cu 1.90	Zn 1.65	Ga 1.81	Ge 2.01	As 2.18	Se 2.55	Br 2.96
Rb 0.82	Sr 0.95	Y 1.22	Zr 1.33	Nb 1.6	Mo 2.16			Rh 2.28	Pd 2.20	Ag 1.93	Cd 1.69	In 1.78	Sn 1.96	Sb 2.05		I 2.66
Cs 0.79	Ba 0.89				W 2.36			Ir 2.20	Pt 2.28	Au 2.54	Hg 2.00	Tl 2.04	Pb 2.33	Bi 2.02		

simple picture of electrons in molecules. MOT is now of increasing importance since many chemists believe it more precisely describes the interaction of atoms forming molecules and the distribution of electrons within them.

Whereas VBT considers electron orbitals of isolated atoms to be conserved in molecules, MOT starts by assuming that a new set of orbitals is created (molecular orbitals) when atoms interact during bond formation. MOT treats the electron distribution in molecules in a similar way to that in which atomic theory treats electrons in atoms. Initially the locations of atomic nuclei are specified. Orbitals (molecular orbitals) are then defined around the nuclei. These molecular orbitals define the region in space around the molecule as a whole where there is a high probability of locating a particular bonding electron. Molecular orbitals are therefore not localised around a single nucleus but extend over part or all of the molecule. The shapes of molecular orbitals are difficult to calculate but a simple approach has been adopted in order to visualise their appearance. The assumption is that molecular orbitals resemble the shape of the atomic orbitals from which they were derived; from the known shape of atomic orbitals the approximate shape of molecular orbitals can be deduced. This method is known as linear combination (addition and subtraction) of atomic orbitals. For any particular pair of atomic orbitals which interact during bond formation, one molecular orbital arises from the addition of parts of the atomic orbital that overlap and one molecular orbital arises due to subtraction of atomic orbitals.

Such principles are illustrated in Figure 1.9, where two atomic s orbitals interact producing a pair of molecular orbitals. The 'bonding' orbital results from the addition of s atomic orbitals in the region of space between the two nuclei. The resulting molecular orbital is of lower energy than either two parent atomic orbitals. The 'antibonding' molecular orbital, which is of greater energy than the parent atomic orbitals, results from subtraction of part of the atomic orbitals that overlap but does not include the region of space between the nuclei. The bonding molecular orbital can be understood more clearly if it is considered to contain electrons which reside in the region between two nuclei and have an energetically favourable interaction with both nuclei. In antibonding molecular orbitals, electrons are influenced by only a single nucleus and will therefore have a less energetically favourable state.

An important point to remember about molecular orbitals is that each pair of atomic orbitals on interacting gives rise to two molecular orbitals, one bonding orbital of lower energy than either parent atomic orbital, and one antibonding orbital of higher energy than either parent atomic orbital.

Combination of s atomic orbitals gives rise to σ (sigma) and σ^* molecular orbitals (antibonding molecular orbitals are signified by an asterisk).

p-Atomic orbitals can interact in two ways, depending on the spatial distribution of the orbitals. If the interacting atomic orbitals lie along the same axis, as is the case for p_x–p_x interactions, then a sigma molecular orbital develops (Figure 1.10). Sigma p molecular orbitals develop in a similar way to that of sigma s, although there is some difference in spatial arrangement of the molecular orbitals since the parent p atomic orbital's geometry differs from that of s atomic orbitals.

A second type of interaction occurs when two atoms have a bond order >1 (i.e. the atoms share two or three pairs of electrons). As shown in Figure 1.11, the additional bonds can form when spatial geometries of two interacting atomic orbitals have a parallel orientation to each other, i.e. p_y–p_y or p_z–p_z interactions. The product is termed a π (pi) molecular orbital in which there is a plane passing through both nuclei along which the electron probability distribution is zero, i.e. the likelihood of finding an electron is negligible. Electrons in pi molecular orbitals reside only above and below the bond axis. It should be noted that two atoms may share all three pairs of p orbitals (i.e. bond order 3) giving rise to one σ and two π bonds.

Figure 1.9. Interaction of s atomic orbitals producing a pair of molecular orbitals.

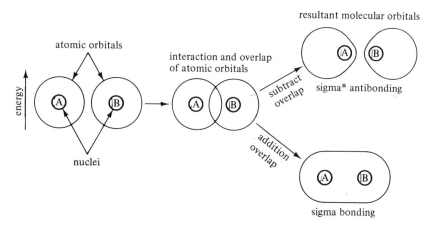

To illustrate how molecular orbitals are used to rationalise the stability of formation of certain molecules and the instability (and therefore unlikelihood of formation) of other molecules from individual atoms, consider the following molecular orbital energy diagrams of

Figure 1.10. Interaction of p atomic orbitals, which lie along the same axis, giving rise to a pair of sigma molecular orbitals.

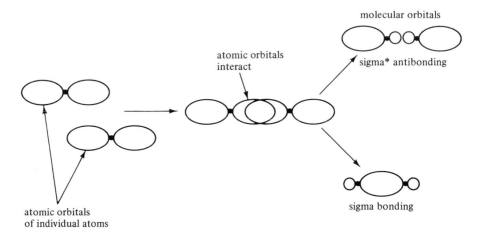

Figure 1.11. Interaction of p atomic orbitals, of two atoms having parallel orientation to each other, giving rise to a pair of π molecular orbitals.

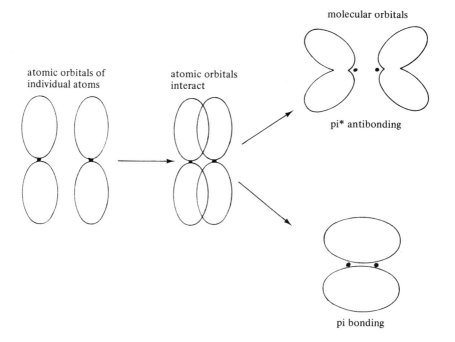

simple molecules (Figures 1.12 and 1.13). Molecular orbitals developed from atomic orbitals of hydrogen atoms are shown in Figure 1.12. The single electron in each hydrogen atom is accommodated in a 1s atomic orbital. When hydrogen atoms come together, the two electrons of the hydrogen molecule both reside in the sigma bonding molecular orbital. Since the electrons now occupy lower energy levels than they did in isolated hydrogen atoms, the hydrogen molecule is more stable than the isolated hydrogen atoms. The difference in energy Δe (Figure 1.12) between atomic orbitals and bonding molecular orbitals is a consequence of the extent of overlap between atomic orbitals in the molecule forming the molecular orbitals. Large overlap produces a large Δe and hence a strong chemical bond. With little atomic orbital overlap the converse is true and a weak bond is formed between component atoms within the molecule.

Helium follows hydrogen in the periodic table and each helium atom has two electrons accommodated in the s subshell. This means that if two helium atoms interacted to form molecular orbitals, two electrons would occupy bonding sigma molecular orbitals whilst the remaining

Figure 1.12. Energy diagram for the formation of molecular orbitals in the hydrogen molecule.

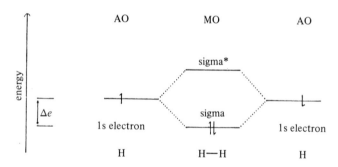

Figure 1.13. Energy diagram describing the formation of molecular orbitals in the hypothetical dihelium molecule.

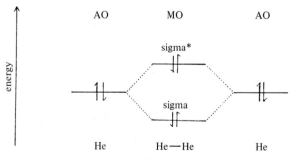

two electrons would have to be located in antibonding sigma orbitals (Figure 1.13). The decrease in energy attributed to bonding orbitals would be cancelled by an increase in energy due to the filling of antibonding orbitals and the binding energy of He_2 would be precisely zero. This argument applies to all inert gases and explains why they are all monatomic.

So far we have considered homoatomic molecules, i.e. molecules composed of atoms of the same element. The energy of each atom's atomic orbitals will be the same. We expect the geometry of resultant molecular orbitals to be symmetrical about a plane perpendicular to and bisecting the inter-nuclear axis. In essence this means that the distribution of negative charge due to electrons in the molecule is also symmetrical. The chemical bond formed under such circumstances is termed a covalent bond. When considering heteroatomic molecules (formed from atoms of different elements) energies of atomic orbitals of each component element are expected to differ from each other. The larger the energy difference between elemental atomic orbitals, the more unsymmetrical are the molecular orbitals which form from them during bond formation. The lack of symmetry in molecular orbitals means the electron distribution is such that the probability of finding particular electrons within the molecule as a whole is more likely in the vicinity of one nucleus than that of another. Molecular orbitals formed between different elements will, in general, not represent true covalent bonds (as in homoatomic molecules), i.e. there will be some degree of polarity in the bond. Generally speaking, the greater the difference in electronegativity of elements, the larger the difference in energy of their respective atomic orbitals. The more electronegative elements will tend to have lower energy atomic orbitals. The difference in energy between atomic orbitals of two elements is a measure of the extent of polarity in the bond. We will consider the nature and type of bond formed between atoms in heteroatomic molecules in more detail in section 1.2.3.

1.2.2 Valency and oxidation number

The term valency describes the number of electrons which an atom utilises in bonding. For example, in methane, CH_4, carbon has four electrons in sp^3 hybrid orbitals (as discussed in Section 1.2.4), each of which forms a sigma bond with a hydrogen atom. The valency exerted by carbon is 4, since the two inner orbital (principal quantum number = 1) electrons are not involved directly in bonding. Each hydrogen atom involves its single electron in bonding, and thus exerts a valency of 1. In carbon tetrachloride, CCl_4, carbon again is sp^3 hybridised and exerts a valency of 4, whilst although each chlorine atom has seven electrons in its outer orbital, only one is shared with carbon in the sigma bond, and chlorine is said to exert a valency of 1.

The rules determining oxidation number (or state) require some explanation. Oxidation number is defined as the charge on the atom in a compound or ion, after bonding electrons have been allocated according to the following general rules:

(i) The oxidation number of hydrogen is +1 (except in metallic hydrides when it is −1 and in hydrogen gas when it is 0).

(ii) The oxidation number of fluorine is always −1 and it never forms more than one bond. Similarly, the oxidation number of all halogens is −1 (except when they are combined with more electronegative elements).

(iii) The oxidation number of oxygen is −2 (except in peroxides when it is −1 and in oxygen gas O_2 when it is 0).

(iv) The oxidation number of an uncombined element is 0.

(v) The oxidation number of monatomic ions is the charge of the ion.

(vi) In any molecule or ion, the sum of the oxidation numbers of all atoms present is equal to the net charge on the molecule or ion.

The following examples explain oxidation numbers in accordance with the above rules. For example in electroneutral CCl_4, chlorine has an oxidation number of −1, there are four of them, hence carbon has an oxidation number of +4. In CH_2Cl_2 each hydrogen has an oxidation number of +1, and each chlorine atom has, an oxidation number of −1. In accordance with rule (vi) carbon has an oxidation number of 0 (zero). Similarly in $SnCl_4$, tin has an oxidation number of +4. In SO_4^{2-} each oxygen atom has an oxidation number of −2. For the overall charge in SO_4^{2-} to be −2, sulphur has an oxidation number of +6. Similarly, in NO_3^-, nitrogen has an oxidation number of +5.

In hydrogen peroxide, H_2O_2, each hydrogen has an oxidation number of +1, hence each oxygen has an oxidation number of −1. In ICl, chlorine, being more electronegative than iodine, has an oxidation number of −1, hence iodine has an oxidation number of +1. In AsH_3 hydrogen has an oxidation number of −1, hence arsenic has an oxidation number of +3.

The oxidation number of an atom in a compound or ion is related to valency, but need not have the same numerical value. Indeed, whilst carbon has a valency of 4 in CCl_4, CH_2Cl_2 and CH_4, its oxidation number in each compound is different. In fact, carbon with its valency of 4 has oxidation number of $+4$ in CCl_4, O in CH_2Cl_2 and -4 in CH_4. Tin and lead can both have a valency of 2 or 4 when their oxidation numbers are $+2$ or $+4$. In $SnCl_2$, tin has an oxidation number of $+2$ and a valency of 2, whilst in $SnCl_4$ it has an oxidation number of $+4$ and a valency of 4.

1.2.3 Types of bond

We have briefly considered how atoms are thought to interact when forming molecules. According to VBT, each atom in a molecule is treated as an isolated atom in which one or more electrons (valence electrons) from one atom are associated with the outer shell of another atom. MOT describes a new set of orbitals developed from the atomic orbitals of interacting atoms forming the molecule. Electrons now belong to the molecule as a whole residing in the new molecular orbitals. Depending on the nature of each element involved in forming a molecule, we recognise that the bond between two atoms can have a varying degree of polarity, i.e. one end of the bond can have a net positive charge, the other end a net negative charge. With that brief resumé of the general principles of bond formation in mind, we may consider particular types of chemical bond.

Covalent bonds

VBT accounts for the non-polar character of covalent bonds by considering that the outer electron (valence electron) from each atom that together form the bond, spend equal lengths of time associated with the outer shell of each atom. MOT says that a non-polar bond arises from the formation of molecular orbitals from the atomic orbitals of the parent atoms which are symmetrical about each nucleus in the molecule. The electrons in the molecule belong to the molecule as a whole, and since the molecular orbitals are symmetrical then the electron distribution is likewise symmetrical.

Good examples of covalent bonds are found in homoatomic molecules, for example H_2, N_2, Cl_2, O_2, etc., and in carbon–carbon bonds. With, for example, Cl_2 molecules, a single covalent bond is formed between two chlorine atoms. If we consider individual nuclei within a Cl_2 molecule (irrespective of which theory of bonding we adhere to) then the outer shell of electrons around each nucleus is considered to have attained the eight electron (octet) configuration of the noble gas argon, a particularly stable arrangement.

Whilst octets usually result during covalent bond formation, there are exceptions. Boron trifluoride (BF_3), for example, has three covalent bonds. Boron itself in BF_3 gains three shared electrons from the fluorines but this still leaves it two electrons short of the octet found in the noble gas neon (Figure 1.14). BF_3 is consequently an extremely reactive compound.

Another important consideration is that atoms when covalently bonded almost invariably have even numbers of s and p electrons. In addition, if covalent bonds are formed by interaction of electrons from different atoms, we assume that the atoms should approach each other in an orientation such that orbitals containing electrons can overlap with each other. As a consequence, when one atom forms several covalent bonds with other atoms within a molecule, the angle between bonds will be influenced by the orientation of the electron filled orbitals. We will consider this again in later sections on hybridisation.

Ionic bonds

Whilst in covalent bonds electrons of individual atoms in a molecule are considered to be shared between atoms, ionic bonds (often called electrovalent bonds) form when one or more electrons from one atom are completely transferred to a second atom. Consider sodium chloride; loss of the single electron in the outer 3s subshell will leave sodium with the electronic configuration of the noble gas neon. Chlorine, by recruiting a single electron to its 3p subshell, will attain the noble gas electronic configuration of argon. As a consequence, sodium will gain a positive charge whilst chlorine gains a negative charge. The resulting charged species (called ions – a positive ion is called a cation whilst the negative ion is referred to as an anion) will attract each other by electrostatic forces (Figure 1.15). Thus the compound, sodium chloride, is formed in which both sodium and chloride ions have attained noble gas electronic configurations (with their concomitant stability) and are held together by an ionic or electrovalent bond.

Figure 1.14. Electron sharing in boron trifluoride.

$$F \!:\! B \!:\! F$$
$$F$$

It is possible for an atom to lose or gain more than one electron during the formation of ionic bonds. Consider aluminium, which has an electronic configuration $1s^2 2s^2 2p^6 3s^2 3p^1$, in which three electrons can be lost giving $1s^2 2s^2 2p^6$ (neon configuration) in the Al^{3+} cation. These three electrons can be donated to three chlorine atoms, one to each atom, forming three chloride anions, thereby creating three ionic bonds between the three Cl^- ions and the single Al^{3+}. In the case of aluminium chloride it is believed the compound exists as a dimer, that is two $AlCl_3$ units combined together forming Al_2Cl_6.

Whilst many ionic compounds contain ions having attained noble gas electronic configurations, there are situations where this is not the case, particularly when transition metals are involved. These elements form cations which have varying numbers of d electrons in their outer shell although they all have complete s and p subshells with noble gas electronic configurations. Instances of ions having odd numbers of s and p electrons are unknown, although odd numbers of d electrons are often encountered (i.e. Mn^{2+}, Fe^{3+}).

Two forms of bonding have been described, namely ionic and covalent. Each describes a set of conditions which in essence are extremes in a continuum. Between classical covalent and ionic bonds are bonds which have varying degrees of covalent or ionic character. The character of a bond between particular elements in heteroatomic molecules is likely to lie somewhere between the two extremes and exhibit some features of each.

Factors favouring ionic bond formation

A bond is likely to have a highly ionic character if:

(1) the metal atom (elements mainly to the left of the period table) forms a cation in which the energy required to remove the electron(s) – the ionisation potential – is low,

(2) the atomic radius of the metal is large since this favours a low ionisation potential,

Figure 1.15. Electrostatic attraction between the sodium cation and chloride anion in sodium chloride. Outer electrons in each ion are shown.

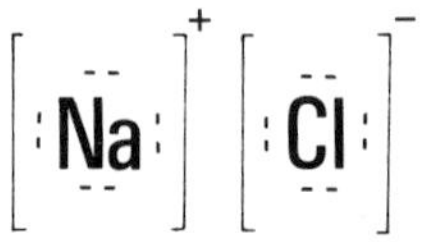

(3) after ionisation, the metal cation attains a noble gas configuration,

(4) the non-metallic atom is small and hence of high electron affinity. Gain of electron(s) by non-metals forming anions is therefore likely to be exoergic (releasing energy) or only weakly endoergic (consuming energy or energy requiring).

Factors favouring covalency, naturally enough, are opposite to those which favour ionic bond formation. With the exception of homoatomic molecules, there will always be a slight difference in electronegativity between different elements forming a chemical bond with a consequent difference in energy of the respective elements' atomic orbitals. Resulting molecular orbitals will, in these circumstances, be unsymmetrical, thereby conferring some degree of polarity to the bond, i.e. some ionic character.

Coordinate bonds

These are also referred to as a dative covalent bonds and from most points of view are identical to an ordinary covalent bond. Coordinate bonds are usually found in coordination compounds (also termed complexes). It should be clear by now that all bonds (covalent and ionic) considered so far have involved pairs of electrons, each atom participating in bond formation supplying one electron to each bond. In coordinate bonds, both bonding electrons are provided (i.e. donated) by the same atom. It is usually the case that the atom which provides the two electrons (and thereby acquires a formal positive charge) is a strongly electronegative atom such as oxygen or nitrogen. As a consequence, the electrons are likely to be held closely to the donor atom thus preventing a large charge separation. An illustration of a coordinate bond is given in Figure 1.16, where the ammonium ion, NH_4^+, is formed (electron pairs in the outermost shell are shown as dots).

Figure 1.16. Ammonium ion as an example of a coordinate bond.

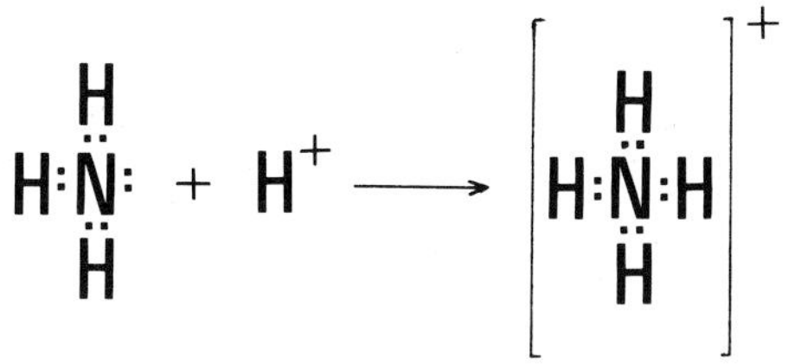

Metallic bonds

The final bond type considered here occurs in metals. Solid metals can be visualised as consisting of an array of cations held together by negatively charged electrons which are free to move throughout the lattice. Charges balance since electrons originate from the neutral metal atoms. Repulsive forces between cations are balanced by attractive forces between individual cations and the cloud of negatively charged electrons.

1.2.4 Molecular geometry and hybridisation

Molecules composed of only two atoms must be linear. With three or more atoms in a molecule there are a number of possible spatial arrangements. According to MOT, for a bond to form between atoms, atomic orbitals must overlap. Each atomic orbital forming the bond has a definite orientation with respect to other orbitals. As a consequence, three or more atoms aligning themselves such that their respective atomic orbitals are correctly positioned in space to overlap with each other will confer a specific three-dimensional configuration to the resultant molecule.

Hybrid orbitals – s and p hybrids

We have seen (Section 1.1.5) the spatial orientation of s and p orbitals. These individual orbitals are identifiable in free atoms of elements. It is possible to merge these atomic orbitals into a new set of orbitals during bond formation. s and p orbitals having merged are termed hybrid orbitals and are designated sp^3 hybrids. The superscript refers not to electron population but indicates that one s and three p orbitals have merged into four new hybrid orbitals, all of which are equivalent energetically. To minimise repulsive forces between electrons, the new hybrid orbitals are spatially orientated towards the four corners of a regular tetrahedron (Figure 1.17).

sp^3 Hybrid orbitals are invoked in understanding the observed shapes of many types of compound. The chemistry of carbon, the enormous variety of carbon compounds known, their structure and the mechanisms whereby they react together are understood and rationalised in terms of hybrid orbitals. Methane (Figure 1.18), for example, has H—C—H bond angles of $109°28'$. The three-dimensional structure of methane is thought of as a central carbon atom with four hybrid orbitals directed towards the corners of a tetrahedron. Electron sharing between sp^3 orbitals and the 1s orbitals of the hydrogens leads to the observed tetrahedral shape. Figure 1.18 is an attempt to represent the three-dimensional shape of methane on a two-dimensional page. Two hybrid bonds (solid lines) are directed along the plane of this page, whilst a third bond (shown as a wedge) projects out of the page and the fourth bond (dotted line) is directed into the page. More complex molecules can be described in terms of sp^3 hybrids, for example ethane, propane and butane. These molecules consist of chains of carbon atoms each being sp^3 hybridised such that bonding is directed towards the corners of a tetrahedron (Figure 1.19). Diamond is a three-dimensional array of sp^3 hybridised carbon atoms.

Hybrid tetrahedral orbitals can account for shapes of molecules which do not contain carbon. For example, the ammonia molecule can

Figure 1.17. Spatial orientation of sp^3 hybrid orbitals.

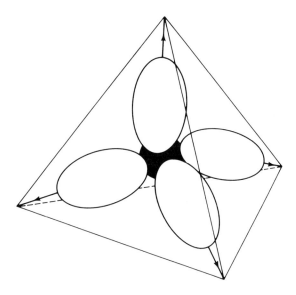

Figure 1.18. Three-dimensional representation of methane.

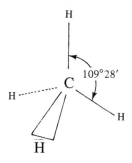

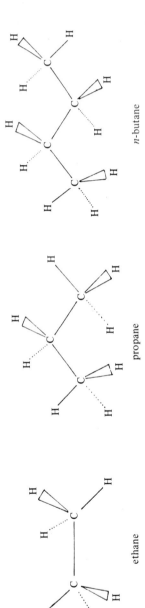

Figure 1.19. Three-dimensional representation of hydrocarbons.

be thought of as a nitrogen atom with five outer electrons distributed between four equivalent tetrahedral hybrid orbitals such that two electrons are paired and occupy one orbital whilst the other three electrons are shared with hydrogen in the remaining three orbitals. This is represented in Figure 1.20 where the lone pair of electrons and one hybrid orbital lie in the plane of the page. The other two hybrids are directed into and out of the page.

Mixing a single s and p orbital gives two sp hybrids with a bond angle of 180°. An example of sp hybrids is found in the compound beryllium chloride ($BeCl_2$). The isolated beryllium atom has a configuration of $1s^2 2s^2$. One of the 2s electrons is promoted to a 2p orbital (resulting in a $1s^2 2s2p$ configuration), thereby forming two sp hybrid orbitals. The electron in each hybrid will be shared with a 3p electron of a chlorine atom forming a bond. The angle between the two hybrids is 180° accounting for the observed linearity of $BeCl_2$.

An example of an sp^2 hybridisation is found in ethene. Reference to Figure 1.21 shows this molecule has a planar arrangement of atoms and

Figure 1.20. Hybrid tetrahedral orbitals in the ammonia molecule.

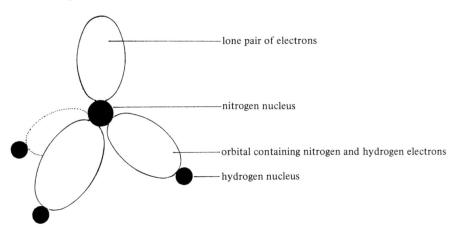

lone pair of electrons

nitrogen nucleus

orbital containing nitrogen and hydrogen electrons

hydrogen nucleus

Figure 1.21. Planar arrangement of atoms in ethene.

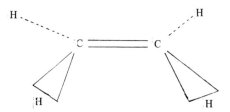

has a double bond between carbon atoms. Mixing one s and two p orbitals gives three hybrids, all lying in a single plane, at an angle of 120° to each other. With two carbon atoms, there are six available hybrid orbitals and two unhybridised p orbitals. The four C—H bonds account for four hybrids. The two remaining hybrids overlap between carbon atoms forming the sigma C—C bond. The pi bond forms by side to side overlap of the unhybridised p_z orbitals (refer to Section 1.2.1).

Hybrid orbitals – involving d orbitals

Amongst the heavier elements, and in particular the transition metals, hybridisation involving a mixture of s, p and d orbitals is common. A few examples of types of hybrid orbitals formed and complexes in which they occur (Figure 1.22) are given below.

Octahedral
Mixing a single s and three p orbitals with the $d_{x^2-y^2}$ and d_{z^2} orbitals gives a set of six equivalent sp^3d^2 hybrids with the orbital orientation directed towards the corners of an octahedron.

Square planar
These are formed through mixing an s, two p orbitals (p_x and p_y) with a $d_{x^2-y^2}$ orbital. The four resulting sp^2d hybrids are directed towards the corners of a square orientated in the xy plane.

Trigonal bipyramidal
One s, three p and the d_{z^2} orbitals combine forming hybrids which are directed towards the five corners of a trigonal bipyramid. In this case the hybrids are not truly equivalent. The three hybrids in the equatorial plane are equivalent to each other. These are formed from two of the p orbitals (p_x and p_y) and the s orbital. The two axial hybrids (which are equivalent to each other) are slightly different from the equatorial hybrids since they are formed from mixing the remaining p_z and d_{z^2} orbitals. Truly equivalent hybrids are formed from equal mixing of all the atomic orbitals utilised to form the hybrids. This does not occur in trigonal bipyramidal hybridisation, hence equivalence only occurs in either the equatorial or axial plane.

Tetrahedral
Tetrahedral hybrids have been encountered previously in, for example, carbon where mixing of s and p orbitals gives four equivalent sp^3

orbitals. Hybrids with tetrahedral symmetry are also formed by mixing an s orbital with three d orbitals (d_{xy}, d_{xz} and d_{yz}).

1.2.5 Crystal field theory

A major feature of the chemistry of transition metals (defined in Section 3.6) is their tendency to form complexes. A complex may be

Figure 1.22. Various hybrid orbitals and complexes in which they occur.

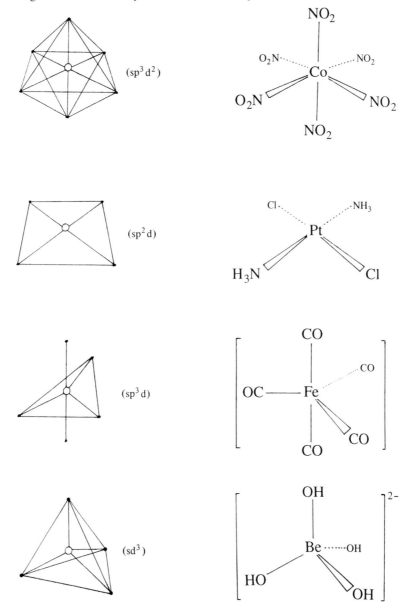

considered as consisting of a central metal atom or ion surrounded by a number of ligands. A ligand can be a single anion, for example Cl^-, F^-, or a group of atoms associated together, for example neutral molecules such as H_2O, NH_3 or charged species such as CN^- (cyanide ion). The interaction of these ligands with the central transition metal or ion is the subject of crystal field theory. This theory, established in 1929, treats the interaction of metal ion and ligand as a purely electrostatic phenomenon where the ligands are considered as point charges in the vicinity of the atomic orbitals of the central atom. This is because, originally, crystal field theory developed through consideration of the way atomic energy levels of ions in crystals were affected by their ionic surroundings. Development and extension of this theory has taken into account the partly covalent nature of bonds between ligand and metal atom, mainly through the application of MOT. Crystal field theory, following these modifications, is often termed ligand field theory.

In essence, the theory describes the splitting of d orbitals of the central metal through interaction of electrons in the d orbitals with electrons associated with the ligand. d-Orbital shapes have been given previously (Section 1.1.5). In brief, there are five d orbitals, two (d_{z^2} and $d_{x^2-y^2}$) are directed along the x, y, z axes whilst three (d_{xy}, d_{zx} and d_{yz}) are directed between the x, y and z axes. All orbitals in the isolated atom are equivalent in energy (i.e. degenerate).

Consider a metal in octahedral coordination with six ligands, each of which will occupy a point at some distance from the central atom along either the x, y or z axis (Figure 1.23) such that all six ligands are

Figure 1.23. Ligand electron clouds as point negative charges in an octahedral environment and their alignment with central metal d orbitals.

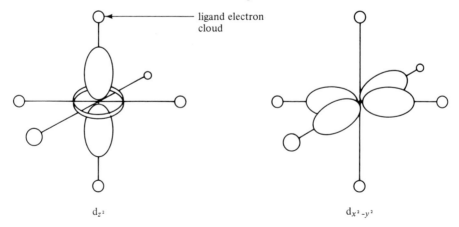

d_{z^2} $d_{x^2-y^2}$

equidistant from the central atom. It is clear from Figure 1.23 that the electron clouds associated with the ligands will be most strongly repelled by electrons in d_{z^2} and $d_{x^2-y^2}$ orbitals since their electron distribution lies along the x, y and z axes, along which the ligands themselves are positioned. Repulsion of ligand electrons by electrons in the remaining d_{xy}, d_{yz} and d_{zx} orbitals will be much less since electron distribution in these orbitals is directed between the x, y and z axes. As a result the d orbitals split into two equivalent sets of orbitals (Figure 1.24). Strong repulsion between ligand and d_{z^2} and $d_{x^2-y^2}$ electrons increases the energy of these d orbitals (referred to as the e_g subset) whilst the energies of d_{xy}, d_{yz} and d_{zx} orbitals (t_{2g} subset) are reduced. The relative energy of these orbitals is shown in Figure 1.24.

Consider a metal residing within a solid crystal lattice (examples of which can be found in Section 5.4), having ligands surrounding it in octahedral symmetry. If we imagine the six ligands approaching the central atom along the x, y, z axes, then the d orbitals will be split into e_g and t_{2g} subsets. d-Orbital electrons of the central atom will then redistribute themselves from whatever d orbitals they were in to the new subsets starting with the low energy t_{2g}. It is also apparent from Figure 1.24 that the t_{2g} subset are lower in energy than the unsplit d orbitals. This is rationalised through the principle of 'preservation of the centre of gravity'. The e_g subset are raised in energy through unfavourable electron interactions, therefore for the algebraic sum of all energy changes to be zero (i.e. return to the energy level of unsplit d orbitals) the t_{2g} subset is invoked. The centre of gravity rule is generally

Figure 1.24. Splitting of d orbitals of a central metal in six coordination due to the interaction with ligands.

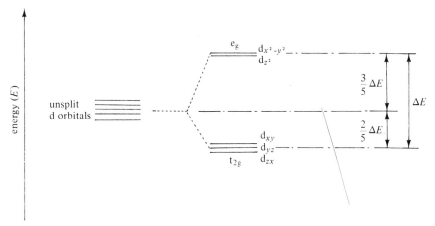

applicable to any splitting pattern when purely electrostatic forces are involved.

d-Orbitals of metals in tetrahedral coordination are also split through the electrostatic field developed by four ligands occupying the apices of the tetrahedron. Figure 1.25 illustrates the orientation of the tetrahedron relative to x, y and z axes and shows the position in space of ligands relative to the central atom. We see that ligand electron density will be directed between x, y and z axes and thereby interact most strongly with d_{xy}, d_{yz} and d_{zx} orbitals which have orientations between x, y and z axes. Energy levels of split d orbitals are given in Figure 1.25. In this case the electron repulsion between ligand electrons and d_{xy}, d_{yz} and d_{zx} orbitals destabilises these orbitals relative to the unsplit condition giving a t_2 subset. The $d_{x^2-y^2}$ and d_{z^2} orbitals are relatively lower in energy and are referred to as the 'e' subset.

1.2.6 Free radicals

In considering bond formation, bond breaking and reactivity, it is important to consider atoms or molecules which are not charged and possess an unpaired electron. Such moieties are called free radicals and are extremely reactive because of their tendency to pair the odd electron with one of opposite spin from another free radical. An example of a free radical which is important in the environment is the hydroxyl radical (OH). Its electron structure may be understood by envisaging the homolytic breakdown of a water molecule to give two neutral species, OH and a hydrogen atom

$$HOH \rightarrow H + OH$$

In breaking the HO—H bond, one electron is transferred to each product, hence producing electroneutral species which are electron

Figure 1.25. Point negative charges in a tetrahedral environment and their effect on the splitting of a d orbital.

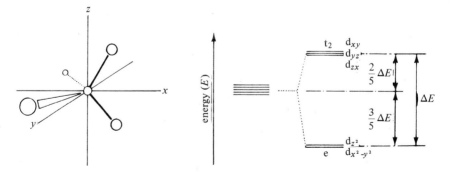

deficient relative to noble gas configuration. The free radical species is thus of high energy and seeks the electron necessary to achieve noble gas configuration. Thus OH will abstract a hydrogen atom from methane, converting itself to the stable molecule H_2O, but creating another reactive free radical, methyl (CH_3)

$$CH_4 + OH \rightarrow CH_3 + H_2O$$

(B)   *Properties of matter*
1.3   Chemical quantities
1.3.1   The concept of a mole of substance

We have seen in Section 1.1.4 how the atom is constituted and how the relative atomic mass of an element is derived using $^{12}_{6}C$ as a reference standard, having assigned to it an atomic mass of 12. Hence the *relative atomic mass* of an element (*atomic weight*) can be defined as the ratio of the average mass per atom, of the natural isotopic composition of an element, to 1/12 of the mass of an atom of $^{12}_{6}C$. Similarly *relative molecular mass* (*molecular weight*) can be defined as the average mass per molecule of the natural isotopic composition of the elements constituting the molecule, to 1/12 of the mass of $^{12}_{6}C$. Both relative atomic and molecular mass values are dimensionless.

In chemical terms the amount of a substance is expressed in moles. One mole of the substance is the amount which contains as many elementary entities as there are atoms in 12 g of $^{12}_{6}C$. This number is called Avogadro's constant and its numerical value is 6.022×10^{23}. Hence there are 6.022×10^{23} atoms in 12 g of $^{12}_{6}C$. When the mole (once known as gram molecule) is used, the elementary units must be specified and may be atoms, molecules, ions, electrons or other specified groups of such particles. There are 6.022×10^{23} atoms in a mole of Hg and $2 \times 6.022 \times 10^{23}$ atoms in a mole of Cl_2. A mole of Hg has a weight of 200.6 g and a mole of Cl_2 a weight of 71.0 g; it follows that a mole of $HgCl_2$ has a weight of 271.6 g.

1.3.2   Chemical equations and stoichiometry

A chemical equation expresses a chemical reaction process, normally showing the reactants on the left side and the products on the right. Thus nitrogen oxide, NO, and ozone, O_3, react to form nitrogen dioxide and molecular oxygen:

$$NO + O_3 \rightarrow NO_2 + O_2 \qquad (1.11)$$

Some processes are highly reversible – dependent upon the specific conditions either the substances on the right side or left side may predominate. These are represented by a double arrow, e.g. the atmospheric equilibrium dissociation of ammonium nitrate, NH_4NO_3 into ammonia, NH_3 and nitric acid, HNO_3:

$$NH_4NO_3 \rightleftharpoons NH_3 + HNO_3$$

In chemical processes, matter can be neither created, nor destroyed. Thus it is usual to make the numbers of atoms of any particular element the same on both sides of the equation, a process known as balancing the equation. Thus the oxidation of nitric oxide by molecular oxygen may be expressed as:

$$NO + O_2 \rightarrow NO_2$$

but more properly the equation is balanced so as to conserve the numbers of both nitrogen and oxygen atoms. Thus:

$$2NO + O_2 \rightarrow 2NO_2$$

Balancing the equation can be useful since it indicates that molecules react usually in simple integral numerical relationships to one another. For the $NO + O_3$ reaction, it is a $1:1$ relationship, whilst for $NO + O_2$ it is $2:1$. These simple numerical relationships are termed the *stoichiometry* of the reaction. Some reactions are apparently non-stoichiometric, i.e. the reactants and products disappear and appear in non-integral ratios. This is not a breakdown of the laws of chemistry, but is usually indicative of a complex mechanism of reaction producing several products, some of which have not been accounted for in the equation.

The concept of molecular weights and stoichiometry allows calculation of masses of reaction products should a reaction proceed to completion.

Worked example
If 1.00 g of nitrogen oxide is completely oxidised in an excess of oxygen, what is the mass of nitrogen dioxide formed?

First write down the balanced equation and the molecular weights of the reactants:

$$2NO + O_2 \rightarrow 2NO_2$$

$$2 \times 30.01 + 32.00 \rightarrow 2 \times 46.01$$

From 1.00 g of NO, the NO_2 produced =

$$1.00 \times \frac{2 \times 46.01}{2 \times 30.01} = 1.53 \text{ g}$$

1.4   States and properties of matter

The three states of matter, solid, liquid and gaseous, are familiar from everyday experience. At the one extreme, in the solid state individual atoms and molecules are in rather fixed positions. They can vibrate about those positions, but do not generally change position relative to one another. Many solids contain a highly ordered array of atoms, ions or molecules and are termed *crystalline*. This crystallinity is often reflected in the regular, angular geometric forms of individual particles, known as crystals. Non-crystalline solids are termed *amorphous*.

At the other end of the scale, gas molecules move with rapid random motion, frequently colliding with one another. Thus mixing is fairly rapid if one gas is introduced into another. Gases can permeate through many solids; for example, water vapour diffuses through polythene sheeting at an appreciable rate. Liquids occupy a position between solids and gases. The molecules do change position relative to one another, but inter-molecular forces are strong and the liquid is more cohesive than the gas. Gases and liquids both have a tendency to flow and are termed *fluids*. Often a solid or a gas may be dissolved in a liquid. In this case the original liquid is termed the *solvent*, and the dissolved substance is the *solute*. In aqueous solutions, ionic compounds tend to dissociate into their constituent ions which become rapidly *solvated*, i.e. they are surrounded by a layer of bonded water molecules. At low concentrations the solvated ions move freely through the liquid independently of one another.

In this section we shall study some of the physico-chemical properties of matter describing basic principles originating in the laboratory valuable in the understanding of environmental and ecological systems.

1.4.1 Laws of gaseous behaviour

For the purposes of environmental chemistry, the gas laws (Boyle's law and Charles's law) may be summarised by the *equation of state of an ideal gas*:

$$\frac{PV}{T} = \text{constant} \tag{1.12}$$

in which P is the pressure, V the volume, and T the absolute temperature in kelvin. Any consistent units of pressure and volume may be used. If specific conditions are defined by pressures P_1, P_2 and P_3 with corresponding volumes and temperatures V_1, V_2 and V_3, and T_1, T_2 and T_3 for our given mass of gas, then

$$\frac{P_1 V_1}{T_1} = \frac{P_2 V_2}{T_2} = \frac{P_3 V_3}{T_3} \tag{1.13}$$

This relationship may be useful in estimating the change in volume of a mass of air when atmospheric temperature and pressure are changed.

Worked example
A sample of air collected at 21°C and 750 mm Hg pressure occupies 1 litre. What would its volume be at standard temperature and pressure? (STP = 273 K and 1 atm; in SI units pressure is expressed in pascals, Pa, where $1\,Pa = 1\,N\,m^{-2}$ and 1 atm = 101 325 Pa = 760 mm Hg = 760 torr = 1.013 bar = 14.70 lb in^{-2}.) The following calculation can be performed using any consistent set of units for pressure.

$$\frac{P_1 V_1}{T_1} = \frac{P_2 V_2}{T_2}$$

$$\frac{750 \times 1}{294} = \frac{760 \times V_2}{273}$$

Thus

$$V_2 = \frac{750 \times 1 \times 273}{760 \times 294} = 0.9161$$

The next obvious question to arise is the nature of the constant in equation (1.12). This may be understood from the *ideal gas law equation*

$$PV = nRT \tag{1.14}$$

in which our constant has been replaced by the term nR, in which n is the number of moles of gas (thus the equation may now be applied to any mass of gas) and R is termed the universal gas constant, and applies irrespective of the chemical composition of the gas under investigation. The value of R may be expressed in a number of units; the most useful in environmental chemistry relate to litres and atmospheres, and therefore

$$R = \frac{PV}{nT} = 0.082056\,\text{l atm K}^{-1}\,\text{mol}^{-1}$$

Taking matters one stage further, having the value of R makes it possible to estimate the volume occupied by one mole of gas at any defined condition of temperature and pressure. The most useful value to remember is at STP, where

$$V = \frac{nRT}{P}$$

$$= \frac{1 \times 0.08205 \times 273.15}{1}$$

$$= 22.41 \, l$$

Thus one mole of any gas at STP occupies 22.41 l. Since this applies to a mole of *any* gas it also applies to mixtures of gases (unless the mixture reacts together chemically to change the total number of moles), including air, a most valuable result for atmospheric chemistry.

In the introduction to this section, the term *ideal gas* was used. This can now be defined as a gas which obeys the above gas laws. In practice, non-ideal behaviour is generally exhibited only by gases close to liquefaction and thus atmospheric air exhibits near ideal behaviour at all times.

1.4.2 Avogadro's law

This law states that equal numbers of molecules of different gases will occupy the same volume, at a given temperature and pressure. The corollary of this statement is that at specific conditions of T and P, equal volumes of gas contain the same number of molecules. In the previous section, we have seen that a mole of any gas occupies 22.41 l at STP and indeed that the gas laws apply irrespective of the chemical composition of the gas. Combining these ideas, it may readily be concluded that, whatever the gas, one mole contains a specific number of molecules. That number, known as the Avogadro constant is 6.022 × 10^{23} mol^{-1}.

1.4.3 Dalton's law of partial pressures

This law applies to mixtures of gases and is thus especially valuable in appreciating atmospheric processes. The partial pressure of each gas in a mixture is defined as the pressure which the gas would exert if it occupied alone the whole volume of the mixture at that temperature. According to Dalton's law, the total pressure of a mixture of gases is equal to the sum of the partial pressures of the constituent gases.

Suppose a vessel of volume v contains three gases in amounts n_1, n_2 and n_3 mol. If each individually occupied the vessel, it would exert a partial pressure of p_1, p_2 and p_3, respectively, at temperature T. The total pressure of gas exerted by the mixture, P, is then given by

$$P = p_1 + p_2 + p_3 \qquad (1.15)$$

If each component of the mixture behaves ideally, then it follows from Section 1.3.1 that

$$p_1 v = n_1 RT$$

$$p_2 v = n_2 RT \qquad (1.16)$$

$$p_3 v = n_3 RT$$

Thus,

$$(p_1 + p_2 + p_3)v = (n_1 + n_2 + n_3)RT$$

Substituting from equation (1.15)

$$Pv = (n_1 + n_2 + n_3)RT = nRT \qquad (1.17)$$

where n is the total number of moles of all gases in the mixture. Dividing equation (1.16) by equation (1.17) it follows that

$$p_1 = \frac{n_1}{n} P$$

and for any gas, i,

$$p_i = \frac{n_i}{n} P = xP$$

The fraction n_i/n is the *mole fraction*, x, of the individual gas, and it follows that the partial pressure of any gas in a mixture may be calculated if the mole fraction and the total pressure of the mixture are known.

Worked example
If air at one atmosphere pressure is considered as a three component mixture comprising 78% nitrogen, 21% oxygen and 1% argon (all by volume), what is the partial pressure of each?

Since equal volumes of gases at a given temperature and pressure contain equal numbers of molecules, it follows that if nitrogen comprises 78% by volume of the mixture, it must also contain 78% of the molecules and likewise of the moles. Nitrogen thus comprises a mole fraction of 0.78 within the mixture.

Thus, for nitrogen,

$$x = 0.78$$

and at $P = 1$ atm,

$$p_{N_2} = 0.78 \times 1$$

$$= 0.78 \text{ atm}$$

Similarly, for oxygen and argon the partial pressures are 0.21 and 0.01 atm.

It is now appropriate to introduce the concept used in air pollution of the *volume mixing ratio*. Suppose in a sample of polluted air, the sulphur dioxide molecules were segregated to one corner of the vessel and their volume measured under the same conditions of T and P as the remaining air. If then, for example, the sulphur dioxide molecules taken from 1 m^3 of polluted air occupied 1 ml,

$$\text{volume mixing ratio} = \frac{1 \text{ ml}}{1 \text{ m}^3} = \frac{1 \text{ ml}}{10^6 \text{ ml}} = 10^{-6}$$

$$= 1 \text{ part per million, or ppm (v/v)}$$

If the sulphur dioxide had instead occupied $1 \mu l$ (i.e. 10^{-3} ml), the volume mixing ratio, or ratio of sulphur dioxide volume to total volume would be

$$\text{volume mixing ratio} = \frac{10^{-3} \text{ ml}}{10^6 \text{ ml}} = 10^{-9} = 1 \text{ part per billion or ppb (v/v)}$$

Now from the example above it will be clear that sulphur dioxide in air at a volume mixing ratio of 1 ppm exerts a partial pressure of 10^{-6} atm if the total pressure of the mixture is 1 atm, and also must represent a mole fraction, x, of 10^{-6}. Thus the volume mixing ratio and the mole fraction are identical. This unit of measurement is especially useful in air pollution since the mole fraction and thus the volume mixing ratio is unaffected by changes in atmospheric temperature and pressure, and thus the conditions of measurement.

1.4.4 Units of measurement of gaseous concentrations

One common means of expression of gaseous concentrations, the mole fraction or volume mixing ratio, is described in the previous section. The other common unit used in atmospheric chemistry is the unit of mass per unit volume, commonly $\mu g\,m^{-3}$ in air pollution. Since the volume of a given air sample, but not the mass of the trace constituent, is affected by changes in T and P, the amount of mass per unit volume will vary with T and P conditions. Interconversion of the two types of unit causes much difficulty but is really extremely simple, as exemplified below.

Worked example
What is a concentration of 1 ppm by volume of sulphur dioxide expressed in $\mu g\,m^{-3}$ at 25°C and 750 mm Hg?

$$1 \text{ ppm } SO_2 \text{ contains 1 ml } SO_2 \text{ per } m^3$$

At STP 1 mol (64.1 g) SO_2 occupies 22.41 l and at 25°C and 750 mm Hg, 1 mol SO_2 occupies

$$= 22.41 \times \frac{298}{273} \times \frac{760}{750} = 24.79\,l$$

Thus under these conditions, 1 ml SO_2 contains

$$64.1 \times \frac{1 \times 10^{-3}}{24.79} = 2.59 \times 10^{-3}\,g$$

This is in 1 m^3 of polluted air, thus

$$SO_2 \text{ concentration} = 2590\,\mu g\,m^{-3}$$

Worked example
What is the concentration of 50.0 $\mu g\,m^{-3}$ of ozone measured at 20°C and 765 mm Hg expressed in ppm by volume?
At STP 1 mol (48.0 g) O_3 occupy 22.41 l, and at 20°C and 765 mm Hg, 48.0 g O_3 occupy

$$= 22.41 \times \frac{293}{273} \times \frac{760}{765} = 23.91\,l$$

Thus 50 $\mu g\,O_3$ will occupy

$$= 23.9 \times \frac{50 \times 10^{-6}}{48.0} = 24.9 \times 10^{-6} \, l$$

$$= 24.9 \times 10^{-9} \, m^{-3}$$

The volume mixing ratio is therefore

$$= 24.9 \times 10^{-9}$$

$$= 24.9 \, \text{ppb}$$

Concentrations of gases and liquids may of course also be described in moles per litre, abbreviated as $mol \, l^{-1}$. A concentration of one $mol \, l^{-1}$ is known as a molar solution, abbreviated as 1M. This terminology is now outdated and not recommended. However it is still widely used and familiarity is necessary.

1.4.5 Diffusion

If a jar of hydrogen and a jar of oxygen are connected end to end, the whole volume will rapidly become uniformly filled by a mixture of the two gases. The process by which they mix is called *diffusion*. In any situation where a concentration gradient exists, diffusive processes will work to remove the gradient. In the case of the jars of hydrogen and oxygen, the diffusion process is driven by the translational (kinetic) energy of the gas molecules which are continuously in motion and colliding with one another. This process is known as *molecular diffusion*.

In the atmosphere, molecular diffusion is too slow to cause appreciable mixing of gases on a large scale, for example car exhaust gases injected into the air. Due to convective mixing processes, the lower atmosphere is frequently highly turbulent and contains many circular eddy motions on scales of centimetres to many metres. These turbulent eddy motions are very effective at mixing the atmosphere, the process being known as *eddy diffusion*. Both molecular and eddy diffusion processes can occur in liquids as well as gases, and pollutants discharged into a river progressively mix with the river water and become diluted due to turbulent eddy diffusion processes.

The generalised diffusion process may be envisaged from Figure 1.26 in which material is diffusing across the imaginary interface $ABCD$ from an area of high concentration to one of low concentration. The flux of material, F, across the interface is defined by Fick's law as

$$F = -D\frac{dc}{dx} \qquad (1.18)$$

in which D is the diffusion coefficient (with units of length^2 time^{-1}) and dc/dx is the concentration gradient. This equation applies to both molecular and eddy diffusion, although the value of D for the two processes is obviously very different. For example, in a room 10 m in length suppose the concentration of CO_2 at one end is 600 mg m^{-3} and at the other end is 700 mg m^{-3}. The concentration gradient is thus 100 mg m^{-3}/10 m or 10 mg m^{-3} m^{-1}. Given that at 273 K the molecular diffusion coefficient of CO_2 is 1.39×10^{-5} m^2 s^{-1}, at that temperature the flux due to molecular diffusion,

$$F = -1.39 \times 10^{-5} \text{ m}^2 \text{ s}^{-1} \times 10 \times 10^{-3} \text{ g m}^{-3} \text{ m}^{-1}$$

$$F = -1.39 \times 10^{-7} \text{ g m}^{-2} \text{ s}^{-1}, \qquad \text{or } -0.14 \,\mu\text{g m}^{-2} \text{ s}^{-1}.$$

The negative sign indicates that the flow of CO_2 proceeds down the concentration gradient, i.e. from high to low concentration.

1.4.6 Osmosis

Osmosis is a special case of diffusion, particularly important in biological systems. Suppose a solution of sucrose (sugar) in water is separated from pure water by a special interface, known as a semi-permeable membrane, which allows passage of water molecules but not of the larger sucrose molecules. There will be a tendency for the water molecules to diffuse from the pure water into the sucrose solution, so as to dilute the sucrose and eliminate the concentration gradient. This process, the spontaneous flow of solvent into a solution, or from a more

Figure 1.26. Molecular diffusion down a concentration gradient.

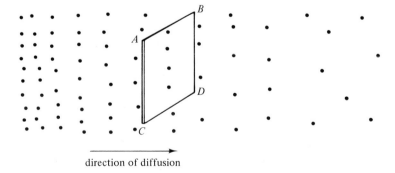

direction of diffusion

dilute to a more concentrated solution, when the two liquids are separated by a suitable membrane is known as *osmosis*.

If the very simple apparatus shown in Figure 1.27 is used, the osmotic tendency will force water into the inverted thistle and up the tube until the hydrostatic pressure of the head of water in the tube matches the osmotic tendency, or pressure, of the water entering the sucrose solution. Now, if sufficient external pressure is applied to the sucrose solution before ingress of water it will stop any inward movement of water; this pressure is known as the *osmotic pressure*. It is defined as the excess pressure which must be applied to a solution to prevent the passage into it of solvent when the two liquids are separated by a perfectly semipermeable membrane.

Plant and animal cells usually contain solutions of salts and sugars enclosed in membranes of a substantially semipermeable nature. Water will then tend to enter the cell setting up an excess pressure inside.

If the variation of osmotic pressure, Π, with temperature T and solution concentration (n/V), is evaluated, it is found that the following relationship holds for dilute solutions:

$$\Pi V = nRT \tag{1.19}$$

This is directly analogous to the equation of state for an ideal gas

Figure 1.27. Simple demonstration of osmotic tendency.

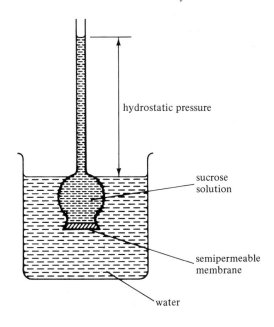

hydrostatic pressure

sucrose solution

semipermeable membrane

water

(Section 1.4.1), the constant R being identical to the gas constant. This is known as the van't Hoff equation. Deviations (or non-ideal behaviour) tend to occur at high concentrations where inter-molecular interactions of solute molecules become important.

1.4.7 Henry's law

In the environment, many situations arise where liquids and gases meet at an interface. Examples include the surface of the ocean, falling raindrops and the interaction of atmospheric gases with the internal fluids of a plant leaf.

Henry's law describes the equilibrium of liquid and gas, and states that the concentration of gas dissolved in a solvent at a given temperature is directly proportional to the pressure of gas in equilibrium with the solution. Thus:

$$c = K_H p \tag{1.20}$$

in which c is the concentration of dissolved gas, p is the partial pressure of gas above the solution, and K_H is the Henry's law constant or partition coefficient.

This relationship of direct proportionality tells us that if the atmospheric abundance, and thus partial pressure, of an unreactive trace gas is doubled, then its equilibrium concentration in surface sea water, for example, should also be doubled. It is important to remember that the solution process is not instantaneous and it will take time to reach a new equilibrium after a change in liquid or gas phase concentration.

A gas which does not react with water is oxygen. Its small but significant solubility in water is of great importance for aquatic life forms, many of which depend upon dissolved oxygen for their existence. The equilibrium water solubility of oxygen is highly temperature dependent, as exemplified by the data in Table 1.3.

One further complication applies to gases which react chemically with the solvent. Consider, for example, carbon dioxide and water. Equation (1.21) expresses the equilibrium of gaseous CO_2 with dis-

Table 1.3. *Concentrations of dissolved oxygen in water in equilibrium with the atmosphere at 760 mm Hg pressure*

Temperature (°C)	0	10	20	30
Dissolved oxygen (mg l^{-1})	14.63	11.28	9.08	7.57

solved CO_2. The latter, however, reacts chemically according to equations (1.22) and (1.23).

$$CO_2(g) \rightleftharpoons CO_2(aq) \tag{1.21}$$

$$CO_2(aq) + H_2O \rightleftharpoons H_2CO_3 \tag{1.22}$$

$$H_2CO_3 \rightleftharpoons 2H^+ + CO_3^{2-} \tag{1.23}$$

These latter processes permit the entry of far more carbon dioxide into solution than might be anticipated from application of Henry's law. The law is valid, when applied to the dissolved species $CO_2(aq)$, but not when applied to all dissolved species derived from CO_2. If equilibrium constants (see Chapter 2) are known for the equilibrium reactions (1.22) and (1.23), then these, combined with knowledge of the Henry's law constant, allow estimation of concentrations of both $CO_2(aq)$ and the other dissolved CO_2-derived species.

Worked examples

(1) Using the data in Table 1.4 the electronic configuration of each element may be deduced as follows:

Na: $1s^2 2s^2 2p^6 3s^1$
Cl: $1s^2 2s^2 2p^6 3s^2 3p^5$
Br: $1s^2 2s^2 2p^6 3s^2 3p^6 3d^{10} 4s^2 4p^5$
Fe: $1s^2 2s^2 2p^6 3s^2 3p^6 3d^6 4s^2$
Ni: $1s^2 2s^2 2p^6 3s^2 3p^6 3d^8 4s^2$
Cu: $1s^2 2s^2 2p^6 3s^2 3p^6 3d^{10} 4s^1$

Table 1.4.

Element	Atomic number (Z)
Na	11
Cl	17
Br	35
Fe	26
Ni	28
Cu	29

(2) The number of protons, neutrons and electrons in

neutral atom $^{14}_{7}N$

negative ion $(^{35}_{17}Cl)^-$

positive ion $({}^{59}_{28}\text{Ni})^{2+}$

neutral atom ${}^{197}_{79}\text{Au}$

negative ion $({}^{37}_{17}\text{Cl})^{-}$

is as follows:

	Protons	Neutrons	Electrons
${}^{14}_{7}\text{N}$	7	7	7
$({}^{35}_{17}\text{Cl})^{-}$	17	18	18
$({}^{59}_{28}\text{Ni})^{2+}$	28	31	26
${}^{197}_{79}\text{Au}$	79	118	79
$({}^{37}_{17}\text{Cl})^{-}$	17	20	18

(3) Given the relative per cent natural abundances of each isotope of the elements (Table 1.5), the mean atomic mass of each element may be calculated as follows:

Table 1.5.

Isotope	Relative abundance (%)	Atomic mass
${}^{12}_{6}\text{C}$	98.89	12.00000
${}^{13}_{6}\text{C}$	1.11	13.00335
${}^{24}_{12}\text{Mg}$	78.99	23.98504
${}^{25}_{12}\text{Mg}$	10.00	24.98584
${}^{26}_{12}\text{Mg}$	11.01	25.98259
${}^{35}_{17}\text{Cl}$	75.77	34.96885
${}^{37}_{17}\text{Cl}$	24.23	36.96699
${}^{112}_{50}\text{Sn}$	1.00	111.9040
${}^{114}_{50}\text{Sn}$	0.70	113.9030
${}^{115}_{50}\text{Sn}$	0.40	114.9035
${}^{116}_{50}\text{Sn}$	14.70	115.9021
${}^{117}_{50}\text{Sn}$	7.70	116.9031
${}^{118}_{50}\text{Sn}$	24.30	117.9013
${}^{119}_{50}\text{Sn}$	8.60	118.9034
${}^{120}_{50}\text{Sn}$	32.40	119.9002
${}^{122}_{50}\text{Sn}$	4.60	121.9034
${}^{124}_{50}\text{Sn}$	5.60	123.9052

mean atomic mass of carbon

$$= 0.9889 \times 12.00000 + 0.0111 \times 13.00335 = 12.0111$$

Similarly,

mean atomic mass of magnesium $= 24.3050$

mean atomic mass of chlorine $= 35.4530$

mean atomic mass of tin $= 118.6847$

Questions

(1) Describe what you understand by the term 'quanta'. Elaborate your answer by reference to the four quantum numbers and the filling of atomic orbitals.

(2) Ionisation potential and electronegativity are important concepts in bonding of elements. Outline the important considerations determining these parameters and how they influence the formation of particular bond types.

(3) Contrast the description of electrons in molecules presented in current theory concerning molecular structure.

(4) How do we account for the trend towards and the stability of metal complexes involving octahedral and tetrahedral coordination?

(5) Describe covalent and ionic bonds in terms of valence bond theory (VBT) and molecular orbital theory (MOT). To what extent will the electronegativity of elements determine the degree of covalency in a bond formed between them? Make reference to bonding in hetero- and homoatomic molecules.

(6) Explain how the varied shape of heteroatomic molecules arises and make reference to the constraints imposed by molecular orbital theory (MOT) on the spatial arrangement of molecular components.

(7) Calculate the number of molecules contained in 1 g of carbon dioxide, 2 g of ozone and $10 \mu g$ of sulphur dioxide.

(8) If $10 \mu g$ of nitrogen oxide reacts with excess ozone to form nitrogen dioxide,

$$NO + O_3 \rightarrow NO_2 + O_2$$

calculate

(i) the mass of nitrogen dioxide formed, and

(ii) the mass of ozone consumed.

(9) Calculate the volume occupied by one mole of an ideal gas at 290 K and 1.02×10^5 Pa. What is the pressure of 1 mol of air at 300 K if it occupies 26 litres?

(10) Express the following volume mixing ratios in units of mass per unit volume at STP:

(i) 340 ppm of CO_2;

(ii) 100 ppb of O_3;

(iii) 10 ppb of SO_2;

(iv) 15 ppb of NO_2.

(11) Express the following mass concentrations (determined at STP) as volume mixing ratios:

(i) $10 \, \mu g \, m^{-3}$ SO_2;

(ii) $2.0 \, mg \, m^{-3}$ CH_4;

(iii) $5 \, ng \, m^{-3}$ $(CH_3)_4Pb$.

(12) Using the data in Table 1.3, calculate the Henry's law constant of molecular oxygen at each temperature, expressed in units of $mol \, l^{-1} \, atm^{-1}$, remembering that 760 mm Hg represents the total pressure of all atmospheric components. Plot a graph of Henry's law constant versus temperature and use it to estimate the solubility of oxygen at 5, 15 and 25°C.

Reference

Allred, A.L. (1961). *J. Inorganic & Nucl. Chem.* **17**, 215–21.

Further reading

Cox, P. A. (1989). *The Elements*: *Their Origin, Abundance and Distribution*. Oxford University Press.

Malone, L.J. (1989). *Basic Concepts of Chemistry*, 3rd edn. New York: John Wiley & Sons.

Segal, B. G. (1989). *Chemistry, Experiment and Theory*, 3rd edn. New York: John Wiley & Sons.

2

Physical chemistry

2.1 Chemical kinetics

2.1.1 Kinetic and thermodynamic control of reactions

A chemical equation can sometimes summarise a process occurring by a number of stages. For example, the reaction

$$2NO + O_2 \rightarrow 2NO_2 \tag{2.1}$$

This reaction could occur in one step by simultaneous collision of two molecules of nitrogen oxide, NO, and one of oxygen. This is statistically improbable, so let us assume that the reaction proceeds by two steps:

$$NO + O_2 \rightarrow NO_3 \tag{2.2}$$

$$NO_3 + NO \rightarrow 2NO_2 \tag{2.3}$$

Each step describes an actual collision of molecules and is known as an *elementary reaction*. The substance NO_3, which is highly reactive and has a very short lifetime (it cannot be isolated from the system), is known as a *reactive intermediate*.

The *law of mass action* applies to elementary reactions, and for reaction (2.2) states that the rate of reaction, $v_{(2.2)}$ is directly proportional to the concentration of both reactants, i.e.

$$v_{(2.2)} = k_{(2.2)}[NO][O_2] \tag{2.4}$$

where $k_{(2.2)}$ is known as the *rate constant* or *rate coefficient* which is a constant peculiar to the reaction and dependent only upon temperature. The square brackets are indicative of molar concentrations of reagents.

All chemical reactions are to some extent reversible. Taking a hypothetical elementary process:

$$aA + bB \rightleftharpoons cC + dD \tag{2.5}$$

Then for the forward and reverse reactions, respectively, the rates,

$$v_f = k_f[A]^a[B]^b \tag{2.6}$$

$$v_r = k_r[C]^c[D]^d \tag{2.7}$$

At equilibrium, the rates of forward and reverse reactions are identical, and

$$v_f = v_r \tag{2.8}$$

$$k_f[A]^a[B]^b = k_r[C]^c[D]^d \tag{2.9}$$

The equilibrium constant, K', for the reaction is defined as

$$K' = \frac{[C]^c[D]^d}{[A]^a[B]^b} = \frac{k_f}{k_r} \tag{2.10}$$

Thus, knowledge of K' allows prediction of the relative proportions of products and reactants in a reaction. For example, returning to reaction (2.1), which may be written as

$$NO + \tfrac{1}{2}O_2 \rightarrow NO_2 \tag{2.11}$$

the equilibrium constant, $K' = 1.38 \times 10^6$ atm$^{-\frac{1}{2}}$ at 25°C. (It is possible to express an equilibrium constant for this reaction process, whether it is an elementary reaction or not.) The units (atm$^{-\frac{1}{2}}$) imply the use of partial pressures, p (which are proportional to molar concentrations; see Chapter 1),

$$K' = \frac{pNO_2}{pNO \cdot pO_2^{\frac{1}{2}}} = 1.38 \times 10^6 \text{ atm}^{-\frac{1}{2}} \tag{2.12}$$

In ambient air, $pO_2 = 0.21$ atm, and

$$\frac{pNO_2}{pNO} = 1.38 \times 10^6 \times (0.21)^{\frac{1}{2}} = 6.3 \times 10^5$$

Thus *at equilibrium* the ratio of NO_2/NO molecules in ambient air should approach one million to one. This is very rarely the case for two reasons:

(a) reactions other than reaction (2.1) affect nitrogen oxide and nitrogen dioxide in the atmosphere. If these are as fast, or faster than, reaction (2.1) they may prevent equilibrium in this reaction from ever being reached.

(b) If the forward and reverse reactions of equation (2.1) are very slow, irrespective of the intervention of other reactions, then it will take an enormously long time to reach equilibrium.

Considering these points, it is an easy matter to gain a crude estimate of the rate of conversion of NO to NO_2 (we look at it this way round since most environmental emissions are of NO). From the law of mass action (assuming reaction (2.1) to be an elementary reaction)

$$\text{rate} = k[NO]^2[O_2] \qquad (2.13)$$

and $k = 14.8 \times 10^3 \, l^2 \, mol^{-2} \, s^{-1}$ at 25°C. Since at STP, 1 mol of pure O_2 occupies 22.4 l, at 25°C,

$$[O_2] = \frac{1}{22.4} \times \frac{273}{298} \times \frac{21}{100}$$

$$= 8.6 \times 10^{-3} \, mol \, l^{-1}$$

for air containing 21% O_2 by volume.

A typical polluted atmospheric concentration of NO is 0.1 ppm (v/v), and

$$[NO] = \frac{1}{22.4} \times \frac{273}{298} \times 0.1 \times 10^{-6}$$

$$= 4.1 \times 10^{-9} \, mol \, l^{-1}$$

Then,

$$\text{rate} = 14.8 \times 10^3 \times (4.1 \times 10^{-9})^2 \times 8.6 \times 10^{-3}$$

$$= 2.1 \times 10^{-15} \, mol \, l^{-1} \, s^{-1}$$

In alternative units,

$$\text{rate} = 2.1 \times 10^{-15} \times 22.4 \times \frac{298}{273} \times \frac{1}{10^{-6}} \times 60 \times 60$$

$$= 1.8 \times 10^{-4} \, ppm \, h^{-1}$$

Thus the initial, maximum rate of conversion of NO to NO_2 is very slow in comparison to the concentration. Note that only the initial rate has been calculated, as the reaction consumes NO, hence reducing the value of [NO] and slowing the reaction. This point is dealt with in later sections.

Why is there an apparent inconsistency between the prediction of the equilibrium constant and observation in the atmosphere? The equilibrium constant represents the situation after infinite time and is ultimately determined by thermodynamics (see Section 2.4). Under atmospheric conditions, true equilibrium is reached in very few reactions, as reaction rates are insufficiently fast. The important reactions in the air tend to be those of greatest speed.

Temperature dependence of rate coefficient: the Arrhenius equation

The rate coefficient is found to vary with temperature so as to obey the equation below, known as the *Arrhenius equation*

$$k = Ae^{-E_a/RT} \tag{2.14}$$

in which A is called the *pre-exponential*, or *frequency factor*, E_a is the *activation energy*, and R and T have their usual meanings. Both A and E_a are constants for a given reaction.

The Arrhenius equation may be explained in terms of simple kinetic theory. E_a represents the energy barrier to a reaction (see Figure 2.1), and $e^{-E_a/RT}$ represents the proportion of molecules possessing energy $>E_a$ and hence able to cross the energy barrier. As T increases, so does $e^{-E_a/RT}$. In a reaction involving bimolecular collisions,

$$A = pZ_{1,2} \tag{2.15}$$

where $Z_{1,2}$ is the collision frequency for molecules of types 1 and 2, and p, which takes values <1 is called the *steric factor* and represents the probability of molecules colliding in the correct orientation for reaction to proceed.

Using this interpretation of a chemical reaction, it is possible to see how thermodynamically (i.e. energetically) favoured products may be formed only very slowly in some reactions. In Figure 2.1, the most thermodynamically favoured reaction products are products (A), but these only form very slowly due to the large energy barrier, E_a, the energy of activation. If the pre-exponential factor for the reaction to products (B) is not vastly different from that for the other reaction, then these products will form more rapidly due to the smaller E_a'. There are some circumstances in chemistry in which a given set of reactants give one product (equivalent to products (B)) at lower temperatures, and another product (equivalent to products (A)) at higher temperatures which enable the larger energy barrier E_a to be overcome. The two products are known, respectively, as the kinetic and thermodynamic reaction products.

In very high temperature environmental processes, such as occur in molten rock, it is normally the thermodynamically predicted product which is formed. This is also true of many ionic reactions in environmental waters. In contrast, as mentioned earlier, in the atmosphere thermodynamic equilibria are rarely attained, and it is kinetics (i.e. reaction rates) which determine which reactions are favoured and what product is formed in most instances.

One interesting point to be considered in more detail later is that a product which is thermodynamically favoured at one temperature may not be favoured at another. Most pollutant nitrogen oxide is formed in combustion processes by combination of molecular nitrogen and oxygen

$$N_2 + O_2 \rightleftharpoons 2NO \qquad (2.16)$$

At the temperature of a flame, thermodynamic equilibrium is approached and very little NO_2 is formed, as NO is the favoured species. At 25°C, as has been shown above, NO_2 is thermodynamically favoured relative to NO, although interconversion may be slow.

Figure 2.1. Energy changes associated with chemical reaction processes.

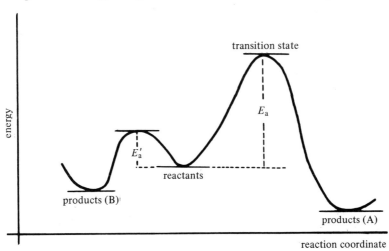

2.1.2 Order and molecularity of reactions

Taking a hypothetical elementary or overall reaction,

$$nA + mB + pC \rightarrow \text{products} \tag{2.17}$$

if, *by experimental observation*,

$$\text{reaction rate} = k[A]^x[B]^y[C]^z \tag{2.18}$$

then the reaction is said to be of *order x* in A, *y* in B and *z* in C, and of overall order $(x + y + z)$. Orders of reaction may be zero or integral, and are occasionally non-integral. Such a *rate equation* may describe an overall reaction containing several steps.

It should be emphasised that, for an elementary reaction, the rate equation may be stated *a priori* from the law of mass action. For example, for atmospheric oxidation of nitrogen oxide by ozone,

$$NO + O_3 \rightarrow NO_2 + O_2 \tag{2.19}$$

$$\text{rate} = k[NO][O_3] \tag{2.20}$$

On the other hand, for an overall reaction containing several elementary steps, experimental determination of the rate equation is the only means of determining reaction orders. For either type of process, the rate coefficient must be determined experimentally.

The *molecularity* of a reaction is the number of molecules in an elementary reaction which come together to form the transition state. Thus, for example, the formation of dinitrogen pentoxide (an important intermediate in the atmospheric formation of nitric acid) is bimolecular:

$$NO_2 + NO_3 \rightarrow N_2O_5 \tag{2.21}$$

This reaction is substantially reversible, the reverse reaction being unimolecular

$$N_2O_5 \rightarrow NO_3 + NO_2 \tag{2.22}$$

Reaction order and molecularity are related only for elementary reactions; for these reactions molecularity and overall order are numerically the same.

2.1.3 Analysis of complex reactions: the steady state principle

Look again at the reaction

$$2NO + O_2 \rightarrow 2NO_2 \tag{2.23}$$

Experimentally, the rate equation is found to be:

$$\text{rate} = k_{\text{exp}}[\text{NO}]^2[\text{O}_2] \tag{2.24}$$

This equation is consistent with a trimolecular elementary reaction. Trimolecular collisions of the appropriate molecules are however statistically improbable, and the relatively high speed of the reaction suggests that this is not a one step process. As mentioned earlier, a two step process involving a reaction intermediate is possible. The stability of that intermediate*, NO_3, is low and hence the first step is markedly reversible:

$$\text{NO} + \text{O}_2 \underset{k_{-1}}{\overset{k_1}{\rightleftharpoons}} \text{NO}_3 \tag{2.2}$$

$$\text{NO}_3 + \text{NO} \underset{k_2}{\rightarrow} 2\text{NO}_2 \tag{2.3}$$

It is possible to test whether this reaction scheme is consistent with the experimentally observed rate equation. From equation (2.3):

$$\text{rate} = \frac{d}{dt}[\text{NO}_2] = 2k_2[\text{NO}_3][\text{NO}] \tag{2.25}$$

Thus the rate of formation of NO_2 is a function of $[\text{NO}_3]$, which is unknown. It is however possible to write an equation for the rate of change of $[\text{NO}_3]$:

$$\frac{d}{dt}[\text{NO}_3] = k_1[\text{NO}][\text{O}_2] - k_{-1}[\text{NO}_3] - k_2[\text{NO}_3][\text{NO}] \tag{2.26}$$

This is seemingly insoluble, as the rate of change of $[\text{NO}_3]$ is dependent upon its concentration. This apparent impasse can only be solved by invoking the *steady state* or *stationary state* principle, which states that reactive intermediates such as NO_3 achieve concentrations which are very small and are essentially unchanged with time, i.e.

$$\frac{d}{dt}[\text{NO}_3] = 0 \tag{2.27}$$

Then

$$[\text{NO}_3] = \frac{k_1[\text{NO}][\text{O}_2]}{k_{-1} + k_2[\text{NO}]} \tag{2.28}$$

Substituting into equation (2.25)

$$\frac{d}{dt}[\text{NO}_2] = \frac{2k_2[\text{NO}] \cdot k_1[\text{NO}][\text{O}_2]}{k_{-1} + k_2[\text{NO}]} \tag{2.29}$$

* In this reaction, NO_3 is the peroxynitrite radical, OONO, not to be confused with the symmetric nitrate radical involved in reactions 2.21, 2.22 and 5.42.

Making the assumption that $k_{-1} \gg k_2[NO]$, then

$$rate = \frac{2k_1k_2}{k_{-1}}[NO]^2[O_2] \qquad (2.30)$$

This is entirely consistent with the experimentally observed rate equation, if

$$k_{exp} = \frac{2k_1k_2}{k_{-1}} \qquad (2.31)$$

In environmental chemistry the steady state principle is often used in a rather different way. In atmospheric photochemistry, there may be tens, or even hundreds of reactions relevant to the formation of a given product. These will include side reactions of intermediates, and other processes causing formation of important reactive intermediates. In order to derive an expression for the formation of a particular product it is necessary to solve a series of differential equations, each expressing the rate of a particular reaction. This may be facilitated if the steady state principle can be applied to the reactive intermediates.

2.1.4 Analysis of first, second and zero order reactions
First order

Taking a hypothetical first order reaction,

$$A \rightarrow B + C \qquad (2.32)$$

Let the initial concentration of A be a mol l^{-1}, and suppose that, after time t, x mol l^{-1} of A have decomposed leaving $(a - x)$ mol l^{-1} of A, and forming x mol l^{-1} of both B and C. Following the earlier conventions,

$$rate = k_1[A] \qquad (2.33)$$

Now, the concentration [A] diminishes with time, and hence

$$rate = \frac{dx}{dt} = k_1(a - x) \qquad (2.34)$$

Separation of variables and integration gives

$$\frac{dx}{(a - x)} = k_1 \, dt \qquad (2.35)$$

$$-\ln (a - x) = k_1 t + \text{const} \qquad (2.36)$$

Since $x = 0$ when $t = 0$, then const $= -\ln a$ and

$$\ln \frac{a}{(a-x)} = k_1 t \qquad (2.37)$$

or

$$x = a(1 - e^{-k_1 t}) \qquad (2.38)$$

It is thus possible to test for first order kinetics by plotting measured values of $\ln [a/(a-x)]$ versus t. A linear plot through the origin (Figure 2.2a) indicates first order kinetics, the slope giving the value of k_1. Figure 2.2(b) indicates the change in concentration of A with time.

The *half life*, τ, of a reaction is the time required to reduce the concentration of A to half its initial value. Substituting $x = a/2$ and $t = \tau$ into equation (2.31)

$$\tau = \frac{\ln 2}{k_1} \qquad (2.39)$$

i.e. τ is independent of initial concentration, a, for a first order process.

One example of a first order environmental process is the decay in activity of a radioisotope. Radioisotopes are characterised by their half life which is independent of concentration. Equation (2.39) may be used to estimate levels of radioactivity from non-replenished environmental sources after a given time period (see Section 3.9.2).

Second order

Many elementary environmental chemical reactions are bimolecular and exhibit overall second order kinetics. Using a notation as before, and taking a hypothetical reaction,

$$A + B \rightarrow C + D \qquad (2.40)$$

At $t = 0$

$$[A] = a \text{ mol}^{-1}$$
$$[B] = b \text{ mol l}^{-1}$$

At $t = t$

x mol l^{-1} of A and of B have reacted forming

x mol l^{-1} of C and D.

For a second order rate law,

$$\frac{dx}{dt} = k_2(a - x)(b - x) \tag{2.41}$$

$$\frac{dx}{(a - x)(b - x)} = k_2\, dt \tag{2.42}$$

This may be integrated by breaking the left hand side into partial fractions, and gives

Figure 2.2. First order kinetic plots.

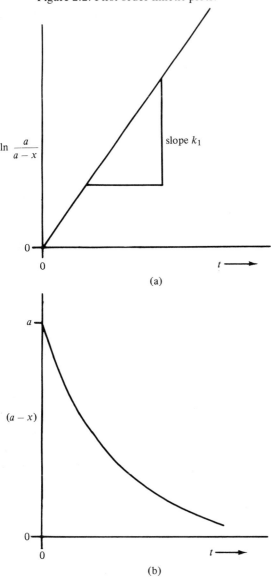

$$\frac{\ln (a - x) - \ln (b - x)}{a - b} = k_2 t + \text{const} \tag{2.43}$$

As $t = 0$ when $x = 0$, const $= \ln (a/b)/(a - b)$

$$\frac{1}{(a - b)} \ln \left(\frac{b(a - x)}{a(b - x)}\right) = k_2 t \tag{2.44}$$

Second order kinetics may be established and k_2 determined by plotting the left hand side of the equation versus t.

For initial concentrations the same, i.e. $a = b$, or for a reaction second order in a single constituent

$$\frac{dx}{dt} = k_2 (a - x)^2 \tag{2.45}$$

$$\frac{1}{(a - x)} = k_2 t + \text{const} \tag{2.46}$$

When $t = 0$, $x = 0$, and thus const $= a^{-1}$

$$\frac{x}{a(a - x)} = k_2 t \tag{2.47}$$

Thus a much simpler rate expression is obtained and a half life may be estimated. Putting $x = a/2$ at $t = \tau$

$$\tau = \frac{1}{k_2 a} \tag{2.48}$$

Zero order

Take the reaction

$$A \rightarrow B \tag{2.49}$$

At $t = 0$, $[A] = a$, and at $t = t$, $[A] = (a - x)$ and $[B] = x$. Then

$$\frac{dx}{dt} = k_0 \tag{2.50}$$

i.e. the rate is independent of $[A]$ for zero order

$$dx = k_0 \, dt \tag{2.51}$$

$$x = k_0 t \tag{2.52}$$

The half life, τ, can be obtained by setting $x = a/2$. Then

$$\tau = \frac{a}{2k_0} \tag{2.53}$$

Thus the half life is directly proportional to initial concentration.

Pseudo first order reactions

Many important atmospheric chemical reactions are pseudo first order. This means that although they are truly second order reactions, the concentration of one component is essentially constant, rendering the reaction rate a function only of the concentration of the other reactant. For example, one of the two important routes of nitric acid formation in air is via reaction of nitrogen dioxide with the hydroxyl radical, OH, the latter being an immensely important reactive intermediate in the air.

$$NO_2 + OH \rightarrow HNO_3 \tag{2.54}$$

Since OH is a reactive intermediate, the steady state principle applies and its concentration may be considered constant over small time intervals (hours, but *not* days). Thus

$$\frac{d}{dt}[HNO_3] = k[NO_2][OH]$$

$$= k'[NO_2] \tag{2.55}$$

where the pseudo first order rate constant,

$$k' = k[OH] \tag{2.56}$$

Then if NO_2 is not replenished by fresh emissions, the concentration at a given time may be estimated from equation (2.37). The appropriate pseudo first order rate constant is given by

$$k' = k[OH]$$
$$= 2.5 \times 10^2 \, ppm^{-1} \, s^{-1} \times 1.0 \times 10^{-7} \, ppm$$
$$= 2.5 \times 10^{-5} \, s^{-1}$$

where $2.5 \times 10^2 \, ppm^{-1} \, s^{-1}$ is the second order rate constant for the NO_2–OH reaction and 1.0×10^{-7} ppm is the concentration of OH.

Then, if the starting concentration of NO_2 is 0.01 ppm (i.e. 10 ppb), after ten hours (36 000 s) the concentration is given by

$$\ln\left(\frac{a}{a-x}\right) = k't \tag{2.37}$$

$$\ln\left(\frac{0.01}{a-x}\right) = 2.5 \times 10^{-5} \times 3.6 \times 10^{4}$$

$$= 9 \times 10^{-1}$$

$$\ln(a-x) = -4.61 - 0.9$$

$$= -5.51$$

$$(a-x) = 4.05 \times 10^{-3} \text{ ppm}$$

$$= 4.05 \text{ ppb}$$

If, however, the NO_2 is replenished by fresh emissions (or more probably by formation from NO emissions) and a concentration of 0.01 ppm is maintained, the formation of HNO_3 in one hour is given by

$$\text{rate} = \frac{d}{dt}[HNO_3] = k'[NO_2]$$

$$= 2.5 \times 10^{-5} \times 3600 \times 0.01 \text{ ppm h}^{-1}$$

$$= 9 \times 10^{-4} \text{ ppm h}^{-1} \text{ or } 0.9 \text{ ppb h}^{-1}$$

Thus, hourly % decay rate for NO_2 due to this reaction is:

$$\% \text{ age decay} = 9 \times 10^{-4} \div 0.01 \times 100$$

$$= 9\% \text{ h}^{-1}$$

i.e. in this case 9% of the total reservoir of NO_2 is consumed each hour, but is continuously replenished by fresh NO_2. HNO_3 is formed at a rate of 0.9 ppb h^{-1}. Such conditions might exist in a polluted urban atmosphere.

2.1.5 Catalysis

Catalysts are substances which alter the rate of a chemical reaction without themselves appearing in the end products. Since it is only the reaction rate which is altered, and not the position of chemical equilibrium, the rates of both forward and reverse reactions in a reversible process must be altered by the same factor (see Section 2.1.1).

Catalysts act by changing the mechanism of a reaction, so as to allow progression via a transition state of lower energy, hence causing a lowering of the activation energy for the reaction (Section 2.1.1).

Catalysis may be all in one phase, termed *homogeneous*, or may occur at the interface of two phases, known as *heterogeneous* catalysis.

Chemical catalysis

An example of homogeneous chemical catalysis is explained in Section 5.1. The reaction between ozone and atomic oxygen in the stratosphere is rather slow and does not affect substantially the concentration of ozone:

$$O + O_3 \rightarrow 2O_2 \tag{2.57}$$

In the presence of nitrogen oxide, however, the process may proceed in two stages, each of which is relatively rapid:

$$O_3 + NO \rightarrow NO_2 + O_2 \tag{2.58}$$
$$NO_2 + O \rightarrow NO + O_2 \tag{2.59}$$

sum $O_3 + O \rightarrow 2O_2$

Hence the catalyst, NO, is changed chemically during the process but returns to its original form without an overall net change.

Heterogeneous catalysts are widely used in industry to accelerate gaseous reactions. The most common catalysts are based upon noble metals (e.g. Pd and Pt) and transition metal compounds (e.g. V_2O_5). Some attention has been given to the catalytic properties of the surface of atmospheric particles in accelerating, for example, the atmospheric oxidation of sulphur dioxide by O_2, O_3 or NO_2. Whilst SO_2 is strongly adsorbed by atmospheric particles and some catalysis of oxidation occurs, this is still a mechanism of minor importance in the atmospheric oxidation of SO_2 (Section 5.1).

A heterogeneous catalytic process used in air pollution control is the employment of noble metals (Pt and Pd) to oxidise carbon monoxide and hydrocarbons in vehicle exhaust gases:

$$2CO + O_2 \rightarrow 2CO_2 \tag{2.60}$$

$$2C_2H_6 + 7O_2 \rightarrow 4CO_2 + 6H_2O \tag{2.61}$$

This oxidation is effective at about 300°C with the catalyst, but would require temperatures of *ca.* 900°C uncatalysed. In auto exhausts, catalysts based upon ruthenium are also used to reverse an oxidation process, i.e. to convert nitrogen oxide into N_2 and O_2:

$$2NO \rightarrow N_2 + O_2 \tag{2.62}$$

As indicated in the previous section, catalysts alter the rate of a reaction, but not the position of equilibrium. Nitrogen oxide is thermodynamically unstable at low temperatures with respect to N_2 and O_2. The breakdown reaction (2.62) is, however, very slow. The catalyst simply speeds progress towards thermodynamic equilibrium by lowering the energy of activation.

Enzyme catalysis

Enzymes are natural proteins which act as highly specific catalysts of biochemical processes. The compound undergoing reaction at the enzyme surface is known as the substrate, and is adsorbed or chemically bonded to the enzyme. Vastly accelerated reaction rates occur in the many enzyme catalyst processes.

There are two major groupings of enzymes. The first group, termed hydrolytic enzymes, accelerate ionic reactions, principally the transfer of hydrogen ions. The other class, termed oxidation–reduction enzymes, catalyse electron transfers.

2.2   Photochemistry

So far, the discussion of chemical reactions has been restricted to thermal processes, i.e. reactions proceeding primarily due to the kinetic energy of the reacting molecules. Photochemical reactions, which are very important in atmospheric chemistry, proceed due to an energy input in the form of light.

Energy associated with light

The energy, E, associated with one quantum of light is given by

$$E = h\nu = \frac{hc}{\lambda} \tag{2.63}$$

where h is the Planck constant, ν is the frequency, λ is the wavelength, and c is the velocity of light. One mole of photons, termed an *einstein*, has energy

$$E_e = Lh\nu \tag{2.64}$$

where L is Avogadro's number. Thus the energy is a function of the wavelength or frequency. The shortest wavelength of solar light penetrating to the troposphere (lower atmosphere) is about 300 nm. For this light

$$E(300 \text{ nm}) = \frac{hc}{\lambda}$$

$$= \frac{6.63 \times 10^{-34} \text{ J s} \times 3.00 \times 10^8 \text{ m s}^{-1}}{300 \times 10^{-9} \text{ m}}$$

$$= 6.63 \times 10^{-19} \text{ J quantum}^{-1}$$

$$= 6.63 \times 10^{-19} \times 6.02 \times 10^{23}$$

$$= 397 \text{ kJ einstein}^{-1}$$

This may be regarded in the context of molecular oxygen dissociation. The energy required for homolytic cleavage of gaseous O_2 into ground state oxygen atoms is termed the *bond dissociation energy*, ΔH.

$$O_2(g) = 2O(g) \quad \Delta H_{298}^{\ominus} = 494 \text{ kJ mol}^{-1} \quad (2.65)$$

Thus, if light of 300 nm is absorbed by molecular oxygen, its energy is insufficient to cause direct chemical change by dissociation. In fact absorption of the light does not occur at this wavelength in any case. In the upper stratosphere (>40 km altitude), shorter wavelengths ($\lambda < 242$ nm) are absorbed with resultant dissociation of molecular oxygen. This reaction leads subsequently to formation of ozone, via

$$O_2 + O + M \rightarrow O_3 + M \quad (2.66)$$

(M is a third molecule, see later), and it is the ozone which is responsible for absorption of light of 242–*ca*. 300 nm in the stratosphere.

It is quite possible for a molecule to absorb light of insufficient energy to cause dissociation. For example, sulphur dioxide absorbs light at *ca*. 384 nm giving rise to an excited state (represented as $*SO_2$)

$$SO_2 + h\nu \rightarrow *SO_2 \quad (\lambda \simeq 384 \text{ nm}) \quad (2.67)$$

Photodissociation does not occur, as the bond dissociation energy of 564 kJ mol^{-1} requires far shorter wavelengths to break the bond. Nonetheless, the formation of an electronically excited SO_2 molecule bearing excess energy may accelerate oxidation of SO_2 via reactions such as (2.68) which are very slow with ground state (unexcited) SO_2

$$*SO_2 + O_2 \rightarrow SO_3 + O \quad (2.68)$$

2.2.1 Photochemical principles

Two basic principles are essential to the appreciation of photo-chemical processes:

(1) Only light that is absorbed by a substance is effective in producing photochemical change.
(2) Each quantum of radiation absorbed by a molecule activates one molecule in the primary step of a photochemical process.

The first principle should be self-explanatory. Unless energy is actually transferred to the molecule by the process of light absorption, no photochemical reaction can ensue. The second principle does not imply that one molecule will react for each quantum absorbed. Commonly, an activated molecule may become deactivated by collision with other molecules without producing chemical change. Alternatively, one activated molecule may set off a chain reaction of many other molecules. This possible degree of variation is expressed by the *quantum yield*, Φ, for a reaction.

$$\Phi = \frac{\text{moles of substance reacted}}{\text{einsteins of light absorbed}} \tag{2.69}$$

2.2.2 A simple photochemical system: NO/NO$_2$/O$_3$

By far the most important photochemical reaction in the lower atmosphere involves photolysis of nitrogen dioxide, which is effective for light of $\lambda \leqslant 435$ nm. The quantum efficiency is approximately one for $\lambda < 370$ nm, and diminishes rapidly as 435 nm is approached (Figure 2.3):

$$NO_2 + h\nu \rightarrow NO + O \tag{2.70}$$

The products are nitric oxide and ground state (unexcited) atomic oxygen (termed O^{3}P)

$$\text{rate} = \frac{-d}{dt}[NO_2] = \Phi I_{abs} \tag{2.71}$$

where I_{abs} is the intensity of absorbed light (einstein l^{-1} s^{-1}). It is worth noting that I_{abs} is a function of the NO$_2$ concentration, and hence at low concentrations little light is absorbed and little reaction occurs. This may be expressed by

$$\frac{-d}{dt}[NO_2] = \Phi k_a[NO_2] = k_1[NO_2]$$

where k_a is a wavelength-dependent absorption constant for NO$_2$, and $k_1 = \Phi k_a$ is a first order rate coefficient. In the atmosphere k_1 is a function not only of quantum efficiency and absorption constant, but of incident light intensity. Thus $k_1 = 0$ at night, and reaches its highest values ($k_1 \simeq 20\,\mathrm{h}^{-1}$) in bright midday summer sunshine.

Moving on from the photodissociation process, the first reaction of the atomic oxygen in the atmosphere is usually with very abundant molecular oxygen

$$O + O_2 + M \rightarrow O_3 + M \qquad (2.66)$$

M is an unreactive third molecule (usually N_2 or O_2) required to absorb excess energy from the reaction. There is a third reaction (2.19) which then completes the cycle:

$$NO_2 + h\nu \xrightarrow{k_1} NO + O \qquad (2.70)$$

$$O + O_2 + \xrightarrow{k_2} O_3 + M \qquad (2.66)$$

$$O_3 + NO \xrightarrow{k_3} NO_2 + O_2 \qquad (2.19)$$

Since the lower atmosphere contains a background of ozone (about

Figure 2.3. Quantum efficiency as a function of wavelength for NO_2 photolysis.

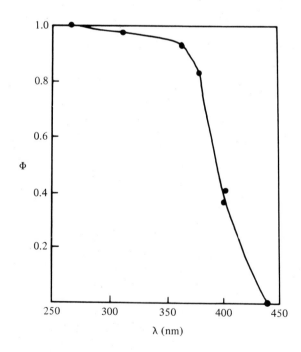

λ (nm)

0.03–0.05 ppm) which has diffused down from the stratosphere, this latter reaction can be very important and does in fact represent a far more rapid means of conversion of NO to NO_2 than the thermal reaction with oxygen discussed earlier.

Each of the three reactions above is rapid, and an equilibrium known as a *photostationary state* is established. This effectively governs the NO_2/NO ratio in air. At equilibrium, NO_2 is formed in reaction (2.19) as rapidly as it is photolysed in reaction (2.70). Thus equating the two reaction rates

$$k_1[NO_2] = k_3[O_3][NO]$$

Hence

$$\frac{[NO_2]}{[NO]} = \frac{k_3[O_3]}{k_1} \tag{2.72}$$

Since k_3 is essentially constant, the NO_2/NO ratio is governed by the ozone concentration and the magnitude of k_1. As k_1 approaches zero (night-time conditions), in the presence of excess ozone NO_2/NO becomes very large, whilst the lowest NO_2/NO ratios are associated with bright sunshine (high k_1). These predictions accord with atmospheric observations and serve to demonstrate the applicability of kinetics rather than thermodynamics in this type of system.

2.3 Thermodynamics

2.3.1 Introduction

The science of thermodynamics evolved from the recognition that the interconversion of heat and work could be predicted from a few relationships. While initial interest concentrated on improving the efficiency of engines, today thermodynamics is an integral part of the study of chemical interactions. Within the scope of thermodynamics one can establish whether or not a reaction is energetically favoured and determine the equilibrium composition of a given suite of chemical constituents (in either case under a prescribed set of conditions). It must be stressed that time-based predictions cannot be made, and hence, kinetic constraints on reactions or the attainment of equilibria cannot be evaluated. Some atmospheric processes which are controlled by kinetic rather than thermodynamic considerations were illustrated in Section 2.1.

In thermodynamic investigations one considers a *system* which is a specifically designated part of the physical world separated from its surroundings by a boundary. In an environmental sense one could study the ocean as a whole or confine one's interest to a clearly designated area such as a fjord or the sea surface microlayer. Systems are categorised as *isolated* when they cannot exchange heat, work or matter with their surroundings, *closed* when energy but not matter may be exported, and *open* when both energy and matter may be exchanged with their environment. Thus, environmental systems are generally open systems but it must be appreciated that the strict application of chemical thermodynamics may be confounded due to biological interactions which render the thermodynamics more complex.

The properties of a system are said to be *extensive* if they depend upon the amount of matter in the system (i.e. volume, mass) and *intensive* when they do not (i.e. density, concentration, temperature). The *state* of the system (that set of circumstances under which all the properties of a system have fixed and reproducible values) may be defined by a limited number of properties known as *state functions* or *state variables*. Thus, the properties of one litre of distilled water at STP will be fixed regardless of the previous history of the water sample.

A *process* refers to any change in the properties of a system. Processes are designated as reversible (ideal) or irreversible (non-ideal). For a *reversible* process, the whole series of changes constituting the process may be reversed such that the original condition may be restored. This process can proceed only in infinitesimal steps such that every step is characterised by a state of balance (i.e. equilibrium). A reversible process necessitates that all sources of energy dissipation (i.e. friction, electrical resistance) are eliminated. This is impossible in reality, and so all actual processes are *irreversible* to some extent. Thermodynamics investigates the change in properties of a system going from an initial state to a final state.

2.3.2 The first law

The total energy (in all possible forms) of a system is termed the *internal energy*, U. Although an absolute value for the internal energy of a system cannot be determined, the change in internal energy associated with the transformation of the system from one state to another can be evaluated. Several forms of energy can be considered. We will limit the following discussion to *work* (w) and *heat* (q) and will deal with electrochemical potential and work associated with surface deformations in later sections.

The work associated with changing the volume of a system is given by the expression:

$$w = -P \cdot \Delta V \qquad (2.73)$$

(P = pressure, ΔV = change in volume.) The sign convention for w is positive when work is done on the system (compression) and negative when work is done by the system (expansion). It should be noted that some texts employ the opposite sign convention, i.e. w is positive when work is done by the system.

The internal energy of a system can also be affected by changes in the temperature. In keeping with our adopted sign convention, q is positive when heat is applied to the system but negative when heat is removed.

The first law of thermodynamics deals with the concept of the conservation of energy and can be expressed in many ways. The total energy of the universe (or any isolated system) is constant. Alternatively, energy cannot be created or destroyed but may be transferred from one form to another. The first law may be stated mathematically as

$$\Delta U = q + w \qquad (2.74)$$

Thus, the internal energy of a system can increase when heat is applied or work is done on the system. Subsequent terms may be added to this equation to account for other types of work, such as electrical or magnetic.

An *adiabatic* process is one during which heat is neither added nor removed from the system (i.e. $q = 0$). If an amount of pressure–volume work w, is performed adiabatically on a system, the change in internal energy is given by

$$\Delta U = w \qquad (2.75)$$

On the other hand, if an amount of heat q is transferred to a system at constant volume (from equation (2.73), $w = 0$ and no mechanical work is done), the change in internal energy is given by:

$$\Delta U = q_V \qquad (2.76)$$

where the subscript V represents the state variable held constant

2.3.3 Enthalpy

Most chemical experiments and environmental processes occur at constant pressure rather than constant volume. Thus, only volume work is performed, and expressions (2.73) and (2.74) can be combined to give

$$\Delta U = q_P - P\Delta V \tag{2.77}$$

If we consider changes in the system from an initial state to a final state (designated by subscripts i and f, respectively) at constant pressure we have

$$U_f - U_i = q_P - P(V_f - V_i) \tag{2.78}$$

and the heat absorbed by the system is therefore:

$$q_P = (U_f + PV_f) - (U_i + PV_i) \tag{2.79}$$

A new state function may be defined by the equation

$$H = U + PV \tag{2.80}$$

such that

$$\Delta H = (U_f + PV_f) - (U_i + PV_i) = q_P \tag{2.81}$$

The state function H is termed *enthalpy*. From expression (2.81) we see that the change in enthalpy for a process equals the heat absorbed by the system if only volume work is performed and the pressure is constant. The state function H is sometimes called the *heat content*, but use of this term may be misleading for processes where the pressure is not constant. Hence, only enthalpy will be used in this text.

2.3.4 Heat capacity

When a system absorbs heat, the temperature increases. The *heat capacity* of a system is defined as the ratio of the change in heat (dq) to the change in temperature (dT). The heat capacity may be determined for both constant volume (C_V) and constant pressure (C_P) systems. Heat capacities for a range of gases are presented in Table 2.1. It follows from expressions (2.76) and (2.81) that at constant volume

$$C_V = \frac{dq_V}{dT} = \left(\frac{\partial U}{\partial T}\right)_V \tag{2.82}$$

and at constant pressure

$$C_P = \frac{dq_P}{dT} = \left(\frac{\partial H}{\partial T}\right)_P \tag{2.83}$$

Temperature changes may also be observed for adiabatic processes. We know from expressions (2.73) and (2.75) that

$$dU = -PdV \qquad (2.84)$$

From equation (2.82) we see further that

$$dU = C_V dT \qquad (2.85)$$

and therefore

$$-PdV = C_V dT. \qquad (2.86)$$

Substituting P from the gas law (Chapter 1) and rearranging, we get:

$$Rd \ln V = -C_V d \ln T \qquad (2.87)$$

Thus, the adiabatic expansion of a gas is associated with a decrease in temperature. This effect is observed in the atmosphere, and dry air cools at the rate of approximately 1°C per 100 m change in elevation. The term *potential temperature*, θ, is used to signify the temperature a parcel of air would attain if moved adiabatically to ground level. In a similar manner, the adiabatic compression of a fluid will cause the temperature to rise. This effect is noticed in deep ocean waters (Table 2.2) where θ refers to the temperature a water body would attain if raised adiabatically to the surface. For ocean waters, the difference between θ and the temperature measured *in situ* is a function of the water temperature, salinity and depth.

Table 2.1. *Heat capacities of gases at 298 K ($J\ K^{-1}\ mol^{-1}$)*

Gas	C_V	C_P
He	12.5	20.8
O_2	21.1	29.4
N_2	20.8	29.1
Cl_2	25.5	33.8
CO	20.9	29.2
CO_2	28.8	37.1
N_2O	30.4	38.7
SO_2	31.6	39.9
NH_3	27.2	35.5

Adapted from Dasent (1982).

2.3.5 Heats of reactions and standard states

Heat may be either absorbed or released from a system during the course of a chemical reaction. The quantity of enthalpy which changes during the reaction is termed the *heat of reaction*. The heat of reaction is temperature dependent and hence the temperature must be specified, usually designated by means of a subscript. In order to compare enthalpy changes for different reactions, the data must be normalised with respect to temperature. The standard temperature most often utilised is 25.00°C (298.15 K). Accordingly, the heat of reaction is expressed as $\Delta H^{\ominus}_{298}$, where the subscript refers to a temperature of 298.15 K and the superscript, $\ominus$, refers to the constituents involved in the reaction being in their standard state. The standard state of reference is 25°C and 1 atmosphere pressure. The *standard state* of an element or compound is the physical state and most stable form which exists under these prescribed conditions. This is considered further in the following section.

When heat is evolved the reaction is said to be *exothermic* and ΔH is negative. A typical example would be the combustion or oxidation of carbon monoxide to carbon dioxide:

$$CO(g) + \tfrac{1}{2}O_2(g) \rightarrow CO_2(g) \qquad (2.88)$$

$$\Delta H^{\ominus}_{298} = -283.0 \text{ kJ}$$

Table 2.2. *Temperature and potential temperatures in the Mindanao Trench*

Depth (m)	Temperature (°C)	θ (°C)
2470	1.82	1.65
2970	1.66	1.44
3470	1.58	1.31
3970	1.59	1.26
4450	1.64	1.25
5450	1.78	1.26
6450	1.92	1.25
7450	2.08	1.24
8450	2.23	1.22
10 035	2.48	1.16

Adapted from Sverdrup, Johnson & Fleming (1942).

Conversely, when heat is absorbed the process is *endothermic* and ΔH is positive. This is the case for the dissociation of water:

$$H_2O(l) = H^+(aq) + OH^-(aq) \qquad (2.89)$$

$$\Delta H^{\ominus}_{298} = 55.5 \text{ kJ}$$

Similarly, enthalpy may increase or decrease when a solute is dissolved in a solvent. The resulting *heat of solution* may be endothermic when heat is absorbed:

$$\text{NaCl:} \quad \Delta H^{\ominus}_{298} = 3.89 \text{ kJ mol}^{-1}$$

or exothermic when heat is evolved:

$$\text{NH}_3\text{:} \quad \Delta H^{\ominus}_{298} = -30.5 \text{ kJ mol}^{-1}$$

As outlined above, enthalpies may be specifically obtained for several other processes such as mixing and phase changes.

2.3.6 Enthalpy of formation

The heat of reaction necessary to form isothermally one mole of a substance from its elementary components is known as the *enthalpy of formation*. Again the customary state of reference is 25°C and 1 atmosphere pressure. The standard enthalpy of formation is therefore expressed as $(\Delta H_f^{\ominus})_{298}$. It should be noted that the enthalpy of formation for an element in its most stable elementary form at these conditions (i.e. in its standard state) is arbitrarily taken to be zero. Thus, $(\Delta H_f^{\ominus})_{298} = 0$ for a gaseous molecular oxygen, liquid molecular bromine, graphite (but not diamond) and rhombic (but not monoclinic) sulphur.

From the above discussion it is clear that, for the combustion of graphite, the heat of reaction equals the enthalpy of formation of carbon dioxide:

$$C(\text{graphite}) + O_2(g) \rightarrow CO_2(g) \qquad \Delta H^{\ominus}_{298} = -393.5 \text{ kJ mol}^{-1} \quad (2.90)$$

$$\Delta H_f^{\ominus} = 0 \qquad \Delta H_f^{\ominus} = 0 \quad \Delta H_f^{\ominus} = -393.5 \text{ kJ mol}^{-1}$$

Enthalpies of formation may be used to calculate the enthalpy change of a reaction. If we consider the general case

$$n_1 X_1 + n_2 X_2 \rightarrow n_3 X_3 + n_4 X_4 \qquad (2.91)$$

The heat of reaction equals the sum of the enthalpies of formation of the products minus the sum of the enthalpies of formation of the reactants. This can be expressed as

$$\Delta H^{\ominus} = \Sigma n_i \Delta H^{\ominus}_{f,products\ i} - \Sigma n_j \Delta H^{\ominus}_{f,reactants\ j} \qquad (2.92)$$

Thus, for the burning of natural gas (methane) we have

$$CH_4(g) + 2O_2(g) \rightarrow CO_2(g) + 2H_2O(g) \qquad (2.93)$$

$$\Delta H^{\ominus} = \Delta H^{\ominus}_f(CO_2) + 2\Delta H^{\ominus}_f(H_2O) - \Delta H^{\ominus}_f(CH_4) - 2\Delta H^{\ominus}_f(O_2)$$
$$\Delta H^{\ominus} = (-393.5) + 2(-241.8) - (-74.8) - 2(0)$$
$$\Delta H^{\ominus} = -802.3\ kJ$$

Calculating the heat of a reaction in this manner is in fact a special application of *Hess's law of heat summation*. This law states that, if a reaction proceeds in a stepwise manner, the sum of the heats of reaction for all steps equals the enthalpy change observed when the reaction proceeds directly. Thus, in the simple case where graphite is oxidised, carbon monoxide may be an intermediate product. The enthalpy change per mole of carbon dioxide formation equals that for the combustion of one mole of graphite to carbon monoxide plus the oxidation of one mole of carbon monoxide to carbon dioxide. This is illustrated as follows:

$$C(graphite) + \tfrac{1}{2}O_2(g) \rightarrow CO(g) \quad \Delta H^{\ominus}_{298} = -110.5\ kJ\ mol^{-1}$$
$$CO(g) \quad\quad + \tfrac{1}{2}O_2(g) \rightarrow CO_2(g) \quad \Delta H^{\ominus}_{298} = -283.0\ kJ\ mol^{-1}$$

$$\overline{C(graphite) + O_2(g) \rightarrow CO_2(g) \quad \Delta H^{\ominus}_{298} = -393.5\ kJ\ mol^{-1}} \quad (2.94)$$

The enthalpy of formation is known for many substances (see Table 2.3), and hence the heat of reaction for very many processes can be determined. However, knowing the heat of reaction only will not allow an assessment of whether or not a particular reaction will proceed to an appreciable extent. Further considerations arise from an appreciation of the second law of thermodynamics.

Worked example
What is the enthalpy change for the oxidation of gaseous H_2S? Is the reaction endothermic or exothermic?

$$H_2S(g) + \frac{3}{2}O_2(g) \rightarrow SO_2(g) + H_2O(g)$$

$$\Delta H^{\ominus} = \Sigma n_i \Delta H^{\ominus}_{f,\text{products } i} - \Sigma n_j \Delta H^{\ominus}_{f,\text{reactants } j}$$

$$\Delta H^{\ominus} = \Delta H^{\ominus}_{f,SO_2(g)} + \Delta H^{\ominus}_{f,H_2O(g)} - \Delta H^{\ominus}_{f,H_2S(g)} - \frac{3}{2}\Delta H^{\ominus}_{f,O_2(g)}$$

$$= -296.8 - 241.8 - (-20.63) - \tfrac{3}{2} \times 0$$

$$= -518.0 \text{ kJ}$$

Because $\Delta H < 0$, the reaction is endothermic.

Table 2.3. *Thermodynamic properties of some compounds of environmental interest* (1 atm, 298.15 K)

Substance	State	$\Delta H^{\ominus}_f$ (kJ mol^{-1})	$S^{\ominus}$ (J mol^{-1} K^{-1})	$\Delta G^{\ominus}_f$ (kJ mol^{-1})
H_2	g	0	130.6	0
O_2	g	0	205.0	0
N_2	g	0	191.5	0
CO	g	−110.5	197.6	−137.2
CO_2	g	−393.5	200.8	−394.4
CH_4	g	−74.8	186.2	−50.8
SO_2	g	−296.8	248.1	−300.2
NH_3	g	−46.1	192.3	−16.5
H_2O	g	−241.8	188.7	−228.6
H_2O	l	−258.8	69.9	−237.2
H_2S	g	−20.63	205.7	−33.56
O_3	g	142.7	239	163.2
NO	g	90.4	210.6	86.7
NO_2	g	33.9	240.5	51.8
N_2O_4	g	9.7	304.3	98.3
N_2O	g	81.5	220.0	103.6
$CaSO_4$	anhydrite	−1434.1	106.7	−1321.7
$CaCO_3$	calcite	−1206.9	92.9	−1128.8
$CaCO_3$	aragonite	−1207.1	88.7	−1127.8
$Al_2Si_2(OH)_4$	kaolinite	−4120	203	−3799
FeS_2	pyrite	−171.5	52.9	−160.2
Fe_3O_4	magnetite	−1115.7	146	−1012.6
MnO_2	pyrolusite	−520.0	53	−465.1
SiO_2	α-quartz	−910.9	41.8	−856.7

Based on Alberty & Daniels (1980) and Stumm & Morgan (1981).

2.3.7 The second law

The second law of thermodynamics is concerned with whether or not a chemical or physical process may occur spontaneously. For example, when two gas cylinders containing air at different pressures are connected, the pressure equalises. Furthermore, the gas composition in each container would be similar. That is to say, there would be no differentiation of the gas mixture leading to the enrichment of oxygen or nitrogen in one cylinder. Similarly, when a hot object comes in contact with a cold object, heat is transferred from the former to the latter causing the temperature to equalise. The existence of pressure, composition and temperature differentials within a system is not precluded by the first law of thermodynamics; however, the processes outlined above would be expected to proceed of their own accord. Such examples are known as *natural* or *spontaneous* processes. Conversely, the reverse processes would be considered *unnatural* or *non-spontaneous*. A manifestation of the second law of thermodynamics is that any system left to itself will approach a final state of rest, that is attain equilibrium. The concept of equilibrium will be developed in Section 2.4. While the first law states that the energy of the universe is constant, the second law states that the entropy of the universe is constantly increasing.

2.3.8 Entropy

The *entropy* of a system is a measure of the degree of disorder that exists in that system. The entropy, S, may be defined by the equation

$$dS = dq/dT \qquad (2.95)$$

where dq is the heat change associated with a reversible process. The sign convention is such that, as the system becomes more random or disperse, the entropy increase is positive. As illustrated by the following examples, entropy changes may be caused by altering the temperature or the number of entities in the system.

For a one component system, an increase in temperature manifests itself by the transformation from a well-ordered solid phase through a semi-ordered liquid phase, to a gas phase which exhibits no structure. If the temperature of a gas is further increased, the kinetic energy increases and the system becomes more random. At each stage the entropy of the system has been increased. Alternatively, when an electrolyte is added to water, the charge on the ion interacts with the dipole moment of adjacent water molecules. This induces localised sites of enhanced structure within the solution, consequently resulting in a decrease in entropy of the solvent.

The greater the number of entities in a system, the more disorder the system exhibits. Hence, the entropy change is positive for reactions in which the number of products exceeds the number of reactants. An important manifestation of this is the *chelate effect* in which several sites on a single molecule may concurrently bind with a metal ion, thereby displacing the coordinated water molecules. This is exemplified by EDTA complexation.*

$$Ca(H_2O)_6^{2+} + EDTA^{4-} \rightarrow Ca\text{–}EDTA^{2-} + 6H_2O \qquad (2.96)$$

On the other hand, the entropy decreases for reactions in which the number of entities decreases such as:

$$H_2(g) + \tfrac{1}{2}O_2(g) \rightarrow H_2O(l) \qquad (2.97)$$

For this example, the entropy change is calculated below.

In a fashion analogous to enthalpy, entropies may be defined for phase changes, mixing and formation. Similarly, a standard state of 1 atmosphere and 25.00°C is generally used for the tabulation of the *standard entropy* ($S_{298}^{\ominus}$) for a substance (see Table 2.3). The entropy change during a reaction may be calculated from the difference between the entropy of the products and reactants. For the general case,

$$n_1X_1 + n_2X_2 \rightarrow n_3X_3 + n_4X_4 \qquad (2.98)$$

The entropy change is given as

$$\Delta S^{\ominus} = \Sigma n_i S_{\text{products } i}^{\ominus} - \Sigma n_j S_{\text{reactants } j}^{\ominus} \qquad (2.99)$$

For the example given above (equation 2.97), the entropy change associated with the formation of water from gaseous hydrogen and oxygen would be

---

* EDTA, or ethylene diamine tetraacetate, is a strong synthetic (man-made) complexing agent for metal ions.

$$\Delta S_{298}^{\ominus} = S_{H_2O(l)}^{\ominus} - S_{H_2(g)}^{\ominus} - \tfrac{1}{2}S_{O_2(g)}^{\ominus}$$
$$= 69.9 - 130.6 - \tfrac{1}{2}(205)$$
$$= -163.2 \text{ J mol}^{-1} \text{ K}^{-1}$$

Worked example

What is the entropy change for the oxidation of gaseous H_2S?

$$H_2S(g) + \tfrac{3}{2}O_2(g) \rightarrow SO_2(g) + H_2O(g)$$

$$\Delta S^{\ominus} = \Sigma n_i S_{\text{products } i}^{\ominus} - \Sigma n_j S_{\text{reactants } j}^{\ominus}$$
$$= S_{SO_2(g)}^{\ominus} + S_{H_2O(g)}^{\ominus} - S_{H_2S(g)}^{\ominus} - \tfrac{3}{2}S_{O_2(g)}^{\ominus}$$
$$= 248.1 + 188.7 - 205.7 - \tfrac{3}{2} \times 205.0$$
$$= -76.4 \text{ J mol}^{-1} \text{ K}^{-1}$$

The third law of thermodynamics also pertains to entropy and may be expressed as: *the entropy of perfect crystalline solids at absolute zero (0 K) is zero*. This clearly has no environmental relevance but is included for completeness.

2.3.9 Helmholtz energy and Gibbs energy

Whether or not a process is energetically favoured can be predicted from the first and second laws of thermodynamics. Two state functions have been defined which combine these concepts. For iso-thermal systems of constant composition and at constant volume the *Helmholtz energy* (Helmholtz free energy), A, is given as

$$A = U - TS \qquad (2.100)$$

Alternatively, for isothermal systems of constant composition and at constant pressure (a situation more commonly encountered in the environment) the *Gibbs energy* (Gibbs free energy), G, is defined as

$$G = H - TS \qquad (2.101)$$

Under these conditions the change in Gibbs energy for a process is therefore

$$\Delta G = \Delta H - T\Delta S \qquad (2.102)$$

As with enthalpy and entropy, a standard state may be designated as 1 atmosphere of pressure and 298.15 K, giving

$$\Delta G_{298}^{\ominus} = \Delta H_{298}^{\ominus} - T\Delta S_{298}^{\ominus} \qquad (2.103)$$

The *standard Gibbs energy of formation*, $\Delta G_f^{\ominus}$, for selected compounds of environmental interest are tabulated in Table 2.3. As with standard enthalpies and entropies, these can be used to determine the Gibbs energy change for a reaction, as illustrated below for the combustion of carbon monoxide:

$$CO(g) + \tfrac{1}{2}O_2(g) \rightarrow CO_2(g) \tag{2.104}$$

$$\Delta G^{\ominus} = \Sigma \Delta G^{\ominus}_{f,products\ i} - \Sigma \Delta G^{\ominus}_{f,reactants\ j}$$

$$\Delta G^{\ominus} = -394.4 - (-137.2 + \tfrac{1}{2} \times 0)$$

$$\Delta G^{\ominus} = -257.2\ kJ$$

Reactions for which ΔG is negative are energetically favoured. Systems will tend to approach a minimum energy level, and therefore, the greater the decrease in Gibbs energy, the more likely the process becomes from a thermodynamic point of view. It must be stressed, however, that kinetics may play an important role. Although a reaction might be predicted to be spontaneous from thermodynamics, it could occur at a sufficiently slow rate as to be indiscernible (see Section 2.1.1).

The Gibbs energy change for a reaction depends upon the change in both enthalpy and entropy. Factors that therefore tend to favour a particular process (i.e. result in a negative ΔG) are a decrease in enthalpy (exothermic) and an increase in entropy. Generally the enthalpy term will dominate, thereby influencing both the sign and magnitude of ΔG. Exceptions to this will occur in gas phase reactions when the number of entities may increase significantly during a reaction. Also, reactions which occur at high temperatures may be entropy controlled due to the temperature coefficient in this term. Some dissolution processes are endothermic but are favoured due to the high positive change in entropy associated with them.

Worked example
What is the free energy change at 25°C for the oxidation of gaseous H_2S? Is the reaction driven by changes in enthalpy or entropy?

$$H_2S(g) + \tfrac{3}{2}O_2(g) \rightarrow SO_2(g) + H_2O(g)$$

$$\Delta G^{\ominus} = \Sigma n_i \Delta G^{\ominus}_{f,products\ i} - \Sigma n_j \Delta G^{\ominus}_{f,reactants\ j}$$

$$= \Delta G^{\ominus}_{f,SO_2(g)} + \Delta G^{\ominus}_{f,H_2O(g)} - \Delta G^{\ominus}_{f,H_2S(g)} - \tfrac{3}{2}\Delta G^{\ominus}_{f,O_2}(g)$$

$$= -300.2 - 228.6 - (-33.56) - \tfrac{3}{2} \times 0$$

$$= -495.2\ kJ$$

Table 2.4. *Gibbs energy changes for the oxidation of organic matter using different electron acceptors*

Oxidation process	ΔG (kJ)
(1) Aerobic oxidation	
$CH_2O + O_2 \rightarrow CO_2 + H_2O$	-2871
(2) Nitrate reduction	
$5CH_2O + 4NO_3^- + 4H^+ \rightarrow 5CO_2 + 2N_2 + 7H_2O$	-2424
(3) Sulphate reduction	
$2CH_2O + SO_4^{2-} \rightarrow 2HCO_3^- + H_2S$	-921
(4) Carbohydrate reduction	
$2CH_2O \rightarrow CO_2 + CH_4$	-239

Alternatively,

$$\Delta G_f^{\ominus} = \Delta H_f^{\ominus} - T\Delta S^{\ominus}$$

From previous worked examples we know that

$$\Delta H_f^{\ominus} = -518.0 \text{ kJ mol}^{-1} \quad \text{and} \quad \Delta S^{\ominus} = -76.4 \text{ J mol}^{-1} \text{K}^{-1}$$

So

$$\begin{aligned}
\Delta G_f^{\ominus} &= -518.0 \times 10^3 - 298.15 \times (-76.4) \\
&= -518.0 \times 10^3 + 22.8 \times 10^3 \\
&= -495.2 \times 10^3 \text{ J mol}^{-1} \\
&= -495.2 \text{ kJ mol}^{-1}
\end{aligned}$$

As mentioned above, systems will tend to attain a minimum energy level. Processes for which the Gibbs energy change are positive will not occur spontaneously. Conversely, processes which allow the Gibbs energy to decrease will proceed spontaneously and can be said to be thermodynamically favoured. A series of competing reactions can be compared with respect to ΔG. The reaction with the most negative ΔG would be the most favoured from a thermodynamic point of view. As noted earlier, there may be kinetic constraints involved such that the process occurs at a rate too slow to be appreciated. However, one would expect that the reaction with the greatest decrease in Gibbs energy would be the most likely to occur.

Consider the possible reactions for the oxidation of organic matter, taken to be CH_2O, given in Table 2.4. The largest energy yield is associated with aerobic oxidation and for normal, well-oxygenated conditions in natural waters, this is the predominant oxidative mechanism. Under anoxic (i.e. oxygen-free) conditions, the nitrate is utilised as an oxidant. This source is quickly exhausted in natural waters. For sea water, abundant sulphate allows the sulphate reduction pathway to proceed following nitrate depletion leading to enrichment in sulphide. Fresh waters contain insufficient sulphate for this to be an important consideration. Alternatively, the thermodynamically least favoured reaction given in Table 2.4 proceeds leading to methane generation. This can be observed in the interstitial waters of anoxic sediments underlying rivers and lakes.

2.3.10 Chemical potential and activity

The discussion of Helmholtz and Gibbs energies in Section 2.3.9 stipulated a system of constant composition. This implies that the composition of the system must exert an influence. Clearly, the internal energy of a system will be affected by the amount of material present. In fact, U increases as the amount of material increases. In keeping with chemical practice, the amount of a component present is measured in terms of moles and designated $n_1, n_2, n_3 \ldots n_i$ whereby the subscript defines the component. The composition must be specified, as different components will contribute varying amounts of internal energy to the system.

Each thermodynamic function not only depends upon P, V and T as outlined in the previous section but also the various n_is present. Accordingly, the Gibbs energy equation (2.102) must be expanded to include the effect of these components, and so becomes

$$\Delta G = \Delta H - T\Delta S + \sum_i \mu_i \, dn_i \qquad (2.105)$$

where μ is the chemical potential. The *chemical potential* was defined by Gibbs to be

$$\mu_i = \left(\frac{\partial G}{\partial n_i}\right)_{T,P,n_j} \qquad (2.106)$$

This is a mathematical definition indicating that, at constant temperature, pressure and composition, with the exception of component i the chemical potential is the change in Gibbs energy due to a change in the amount of component i present. Thus, the Gibbs energy is an extensive property, and chemical potential is an intensive property.

The ionic strength (which is related to the total amount of electrolytes dissolved) of a solution affects equilibrium conditions. Observations associated with an increase in the ionic strength include the enhanced dissociation of a weak electrolyte and an increase in the solubility of a given precipitate. These phenomena arise because cations and anions will be mutually attracted in an electrolytic solution. A certain proportion may exist as an ion pair (see Section 2.4.5). This is a transient and weak association between a cation and an anion. Reactions and equilibrium processes involve only the dissociated cations and anions. Hence, the formation of ion pairs essentially decreases the number of ions which may be involved in these other processes. This is described as decreasing the 'effective' concentration. The greater the ionic strength becomes, the more profound the change in the effective concentration. Accordingly, a concentration scale must be defined which is independent of the electrolytic composition such that a unique set of constants may be defined to describe reactions and systems under any ionic strength. This is achieved by introducing the concept of activity, whereby activity (expressed as λ_i or a_i) is the 'effective' concentration of a constituent i in solution. Activity is related to concentration by the expression:

$$a_i = \gamma_i c_i \tag{2.107}$$

where γ_i (sometimes given as f_i) is the activity coefficient of component i and c_i is the concentration of component i. The activity coefficient is conventionally taken to be dimensionless. As a result, the *absolute activity* (distinguished using λ_i notation) has the same units as the concentration term. Several concentration scales can be applied, the commonly utilised being

> molar – moles of solute per litre of solution (c_i)
> molal – moles of solute per kilogram of solvent (m_i)
> mole fraction – the ratio of moles of solute to moles of solvent.
> (Mole fraction, being a ratio, is of course dimensionless.)

The popular alternative to absolute activity is the dimensionless relative activity (a_i). *Relative activity* is the ratio of the absolute activity of component i in solution to the standard state absolute activity of pure component i (i.e. generally 1 mol dm^{-3} or 1 mol kg^{-1}).

The activity coefficient is a function of ionic strength. This relationship is given by the *Debye–Hückel equation*:

$$-\log \gamma_i = \frac{0.51 Z_i^2 \sqrt{I}}{1 + 0.33 a_i \sqrt{I}} \tag{2.108}$$

where a_i, the ion-size parameter, is the effective diameter of the hydrated ion in Ångstrom units ($1\ \text{Å} = 10^{-1}\ \text{nm}$). This is about $3\ \text{Å}$ for most univalent ions but may be as large as $11\ \text{Å}$ for ions exhibiting multiple charges. Regardless of the ions involved, as the ionic strength approaches zero, that is, at infinite dilution, the activity coefficient approaches unity. Hence, the activity and the concentration are the same for very dilute solutions. The ionic strength, I, of a solution can be computed using the expression

$$I = \tfrac{1}{2}\Sigma\, c_i Z_i^2 \tag{2.109}$$

where c_i is the concentration of an individual ion and Z_i is the charge on that ion. The units of I will be the same as that used for c_i.

Worked example
Calculate the ionic strength of (i) 0.7 M NaCl and (ii) 0.7 M Na$_2$CO$_3$.
(i) 0.7 M NaCl

$$\begin{aligned}
I &= \tfrac{1}{2}[c_{\text{Na}^+} \times Z_{\text{Na}^+}^2 + c_{\text{Cl}^-} \times Z_{\text{Cl}^-}^2] \\
&= \tfrac{1}{2}[0.7 \times (1)^2 + 0.7 \times (-1)^2] \\
&= 0.7\ \text{M}
\end{aligned}$$

(ii) 0.7 M Na$_2$CO$_3$

$$\begin{aligned}
I &= \tfrac{1}{2}[c_{\text{Na}^+} \times Z_{\text{Na}^+}^2 + c_{\text{CO}_3^{2-}} \times Z_{\text{CO}_3^{2-}}^2) \\
&= \tfrac{1}{2}[1.4 \times (1)^2 + 0.7 \times (-2)^2] \\
&= 2.1\ \text{M}
\end{aligned}$$

The chemical potential can also be expressed in terms of activity, using

$$\mu_i = \mu_i^{\ominus} + RT \ln a_i \tag{2.110}$$

where $\mu_i^{\ominus}$ is the standard state chemical potential of component i. The *standard state* is usually designated to be the system that exists when component i has unit activity at 1 atmosphere pressure and 25°C. Under these conditions, the second term in equation (2.110) becomes zero. For reasons to be outlined shortly, the conditions for which unit activity hold will be dependent upon the concentration scale utilised. By substituting from equation (2.107), the chemical potential can be expressed in terms of molar concentrations, c_i:

$$\mu_i = \mu_i^{\ominus} + RT \ln \gamma_i c_i \tag{2.111}$$

While molar concentrations are temperature dependent, molal concentrations are independent of temperature. Hence, the chemical potential can alternatively be expressed as

$$\mu_i = \mu_i^* + RT \ln \gamma_i' m_i \tag{2.112}$$

where m_i is the molal concentration of component i, γ_i' is the activity coefficient of component i relating activity to molal concentrations, and μ_i^* is the standard state chemical potential (i.e. again when component i has unit activity but as derived from the molal concentration scale). The chemical potential, μ_i, is an intrinsic property of a system. For any given total amount of a component i in solution, the chemical potential will be constant. However, as evident from expressions (2.111) and (2.112), this total amount of chemical potential can be defined as two discrete components. While the summation of the two terms will be constant, the magnitude of the individual terms can vary. Thus, it should be noted that γ_i' and μ_i^* used in equation (2.112) do differ from γ_i and $\mu_i^\ominus$ as used in equation (2.111). Activity coefficients and standard state chemical potentials are defined by stipulating the reference state and standard state. The activity coefficient is essentially defined (or calibrated) by the choice of the reference state. The *reference state* stipulates the concentration scale in use and sets a limit on the activity coefficient. This limit is generally given as the solution condition for which the absolute activity equals the concentration. Conventionally then γ_i approaches unity as the c_i approaches $0\ \mathrm{mol\ dm^{-3}}$.† Alternatively, γ_i' approaches unity as the m_i approaches $0\ \mathrm{mol\ kg^{-1}}$. The activity coefficients are dimensionless but will vary with the concentration scale in use. Considering the case for a constant activity of component i in solution, clearly c_i is not equal to m_i. Thus, γ_i (i.e. a_i/c_i) cannot equal γ_i' (a_i/m_i).

Although both $\mu_i^\ominus$ and μ_i^* are standard state chemical potentials for component i, these values differ. The standard state was defined previously by the system containing component i at unit activity. The apparent contradiction in these two statements can be clarified in considering further the definition of the standard state. The *standard state* is the ideal or hypothetical system at which component i exhibits unit concentration (i.e. $1\ \mathrm{mol\ dm^{-3}}$ in the molar scale but $1\ \mathrm{mol\ kg^{-1}}$ in the molal scale) *and* the activity coefficient is also unity. Obviously $1\ \mathrm{mol\ dm^{-3}}$ does not equal $1\ \mathrm{mol\ kg^{-1}}$, and so $\mu_i^\ominus$ does not equal μ_i^*.

† Note that $1\ l \equiv 1\ \mathrm{dm^3} \equiv 1000\ \mathrm{ml} \equiv 1000\ \mathrm{cm^3}$.

Worked example
Given that the ion-size parameter for Ca^{2+} is 6 Å, calculate the activity of Ca^{2+} in a solution containing 1.50×10^{-4} mol dm^{-3} $CaCO_3$ and 2.20×10^{-4} mol dm^{-3} Na_2CO_3.

First calculate the ionic strength knowing the ions involved are Na^+, Ca^{2+} and CO_3^{2-}:

$$
\begin{aligned}
I &= \tfrac{1}{2}\Sigma \, c_i Z_i^2 \\
&= \tfrac{1}{2}(2 \times 2.20 \times 10^{-4} \times 1^2 + 1 \times 1.50 \times 10^{-4} \times 2^2 \\
&\quad + 1 \times 1.50 \times 10^{-4} \times 2^2 + 1 \times 2.20 \times 10^{-4} \times 2^2) \\
&= \tfrac{1}{2}(4.40 \times 10^{-4} + 6.00 \times 10^{-4} + 6.00 \times 10^{-4} + 8.80 \times 10^{-4}) \\
&= 1.26 \times 10^{-3} \text{ mol dm}^{-3}
\end{aligned}
$$

Next calculate the activity coefficient for Ca^{2+}:

$$
-\log \gamma_i = \frac{0.51 \times Z_i^2 \sqrt{I}}{1 + 0.33 a_i \sqrt{I}}
$$

$$
-\log \gamma_{Ca^{2+}} = \frac{0.51 \times 4 \times 3.55 \times 10^{-2}}{1 + 0.33 \times 6 \times 3.55 \times 10^{-2}}
$$

$$
= 6.77 \times 10^{-2}
$$

$$
\gamma_{Ca^{2+}} = 0.856
$$

Finally calculate the absolute activity of Ca^{2+}:

$$
\begin{aligned}
a_{Ca^{2+}} &= \gamma_{Ca^{2+}} c_{Ca^{2+}} \\
&= 0.856 \times 1.50 \times 10^{-4} \\
&= 1.28 \times 10^{-4} \text{ mol dm}^{-3}
\end{aligned}
$$

The relative activity is dimensionless and would be

$$
a_{Ca^{2+}} = 1.28 \times 10^{-4}
$$

The discussion of activity above has been concerned only with solutions. Pure solids and liquids, including water, are defined to have an activity of unity. Whereas in solutions the activity represents the idealised or effective concentration, the idealised pressure of a gas is defined as the *fugacity, f*. For a gaseous compound, *i*, therefore

$$
f_i = \gamma_i \times p_i \tag{2.113}
$$

where f_i is the fugacity of gas *i* having a partial pressure p_i, and γ_i is the *fugacity constant.* The fugacity constant approaches unity as the total pressure approaches zero. Finally, chemical potentials may be expressed in terms of fugacity as:

$$\mu_i = \mu_i^\ominus + RT \ln f_i \qquad (2.114)$$

The standard state for gases is taken to be one for which the fugacity, rather than pressure, is unity.

The relationship between chemical potential and activity will be considered further in the following section.

2.4 Equilibria

2.4.1 Introduction

For a system which experiences a set of constant external conditions which completely determines its state, the properties of the system will not change with time and the system is said to exist in a state of rest. When the state of the system is such that the initial state is re-established following the temporary alteration of external conditions, the system is said to exist in a state of *equilibrium*. This must be considered in a macroscopic sense because it is the observable or measurable properties which are invariant with respect to time. Individual particles or molecules can behave in a complicated manner, undergoing phase changes or participating in chemical reactions. Thus, while equilibrium may be a state of rest for the system as a whole, dynamic processes continually occur. However, for any process the forward and reverse rates are comparable ensuring that the overall composition of the system does not alter.

Several criteria can be established that describe a system at equilibrium. The first and second laws of thermodynamics indicate that a system will tend to change to a state of minimum energy and maximum entropy. These conditions must therefore be satisfied for a system to attain equilibrium. A spontaneous process will cause a net positive change in the entropy of a system. Clearly a system at equilibrium can undergo no spontaneous changes. Therefore, the entropy change for any equilibrium process must be zero:

$$dS = 0 \qquad (2.115)$$

Similarly it can be shown that the Gibbs energy change must be zero for any process in a system at constant composition at equilibrium under isothermal and isobaric conditions. Consider the definition of Gibbs energy:

$$G = H - TS$$

The substitution of H with $U + PV$ and differentiation gives:

$$dG = dU + P\,dV + V\,dP - S\,dT - T\,dS \qquad (2.116)$$

We know that

$$dU = dq + dw$$

$$dq = T\,dS$$

and for a system in which only pressure–volume work is considered

$$dw = -P\,dV$$

Therefore, substitution into equation (2.116) gives:

$$dG = T\,dS - P\,dV + P\,dV + V\,dP - S\,dT - T\,dS \qquad (2.117)$$

or

$$dG = V\,dP - S\,dT \qquad (2.118)$$

Thus, the criterion for equilibrium with respect to a process occurring at constant pressure and temperature is

$$dG = 0 \qquad (2.119)$$

Alternatively, it could be imagined that if the system could perform any work leading to inhomogeneities of either temperature or pressure within the system, this would necessarily be followed by a spontaneous process equalising either temperature or pressure.

As indicated above, a fundamental property of a one component system at equilibrium under conditions of constant temperature and pressure is that the Gibbs energy change for any process is zero. This criterion holds similarly when a multi-component system is considered. Thus, from equations (2.105) and (2.119) we get

$$\sum_i \mu_i\,dn_i = 0 \qquad (2.120)$$

If we consider the transfer of a small amount, δn_i moles of a component i from phase α to β, the Gibbs energy change would be given as

$$\delta G = -\mu_i^\alpha \delta n_i + \mu_i^\beta \delta n_i \qquad (2.121)$$

where μ_i^α and μ_i^β represent the chemical potential of component i in phases α and β, respectively. With the conditions that $\Delta G = 0$, we find that at equilibrium:

$$\mu_i^\alpha = \mu_i^\beta \qquad (2.122)$$

In a closed system at equilibrium under conditions of constant pressure and temperature, the chemical potential of compound i is the same for all phases in which i exists.

2.4.2 Phase equilibria

A phase is a part of a system which is homogenous with respect to chemical and intensive physical properties and separated from other parts of the system by boundary surfaces. For example, water containing ice constitutes only two phases. This holds true whether the ice is in one or a number of pieces because each ice fragment will be identical with respect to chemical and intensive physical properties.

Phase changes are of limited importance in understanding the chemistry and distribution of components of environmental interest. The hydrogeological cycling of water necessarily involves solid, liquid and gaseous phases. Calcium carbonate is also known to undergo phase changes in the environment; aragonite and calcite are both pure phases of calcium carbonate. At 1 atmosphere pressure and 25°C, the standard Gibbs energies of formation for these phases are -1127.8 and $-1128.8 \, \text{kJ mol}^{-1}$, respectively. This indicates that calcite is the more stable. Aragonite is known to precipitate from shallow waters in the Bahamas; however, it is slowly transformed into calcite during diagenesis (i.e. the post-depositional alteration of sediments).

2.4.3 Chemical equilibria

An equilibrium constant, K', has already been defined by the *law of mass action* (Section 2.1.1) for the following general reaction:

$$aA + bB \rightleftharpoons cC + dD \qquad (2.123)$$

as

$$K' = \frac{[C]^c[D]^d}{[A]^a[B]^b} \qquad (2.124)$$

However, the equilibrium constants can also be derived thermodynamically. The Gibbs energy for a component i can be expressed in terms of the activity using:

$$G_i = G_i^{\ominus} + RT \ln a_i \qquad (2.125)$$

Therefore, the Gibbs energy for the constituents of the general reaction given above are*:

$$aG_A = aG_A^{\ominus} + aRT \ln \{A\}$$
$$bG_B = bG_B^{\ominus} + bRT \ln \{B\}$$
$$cG_C = cG_C^{\ominus} + cRT \ln \{C\}$$
$$dG_D = dG_D^{\ominus} + dRT \ln \{D\}$$

The Gibbs energy change for this reaction is

$$\Delta G = cG_C + dG_D - aG_A - bG_B \qquad (2.126)$$

$$= cG_C^{\ominus} + dG_D^{\ominus} - aG_A^{\ominus} - bG_B^{\ominus} + RT \ln \frac{\{C\}^c \{D\}^d}{\{A\}^a \{B\}^b} \qquad (2.127)$$

The standard Gibbs energy change for the reaction is defined as

$$\Delta G^{\ominus} = cG_C^{\ominus} + dG_D^{\ominus} - aG_A^{\ominus} - bG_B^{\ominus} \qquad (2.128)$$

For a system at equilibrium, $\Delta G = 0$, leaving the relationship:

$$-\Delta G^{\ominus} = RT \ln \frac{\{C\}^c \{D\}^d}{\{A\}^a \{B\}^b} \qquad (2.129)$$

As $G^{\ominus}$ is constant for a given temperature and pressure, this expression becomes

$$-\Delta G^{\ominus} = RT \ln K \qquad (2.130)$$

where the equilibrium constant, K, is given as:

$$K = \frac{\{C\}^c \{D\}^d}{\{A\}^a \{B\}^b} \qquad (2.131)$$

It should be noted that the equilibrium constant defined here differs from that derived in Section 2.1.1 in that activities rather than concentrations have been used. Clearly K and K' are related as follows:

---

* Square brackets [ ] are used to denote concentrations; curly brackets { } denote activities.

$$K = \frac{\{C\}^c \{D\}^d}{\{A\}^a \{B\}^b}$$

$$= \frac{[C]^c [D]^d}{[A]^a [B]^b} \times \frac{\gamma^c \times \gamma^d}{\gamma^a \times \gamma^b}$$

$$= K' \times \frac{\gamma^c \times \gamma^d}{\gamma^a \times \gamma^b}$$

In most discussions the activity coefficients are assumed to be unity, and so

$$K = K' \tag{2.132}$$

In order to calculate equilibrium constants from tabulated thermodynamic data, equation (2.130) may be rewritten in the form:

$$\log_{10} K = \frac{-\Delta G^{\ominus}}{2.303 RT} \tag{2.133}$$

For example, the equilibrium constant for carbon monoxide combustion (equation 2.104) is calculated as:

$$\log_{10} K = \frac{-(-257.2) \times 10^3}{2.303 \times 8.314 \times 298.15}$$

$$= 45.05$$

$$K = 1.1 \times 10^{45} \text{ atm}^{-\frac{1}{2}}$$

(R is taken as $8.314 \text{ J K}^{-1} \text{ mol}^{-1}$.)

Le Chatelier's principle states that, if a system at equilibrium is perturbed, the system will react in such a way as to minimise the effect of the perturbation. Consider the general case for a system containing A and B. The reaction producing C and D proceeds until equilibrium is established. The equilibrium concentrations of the reactants and products are determined by K. Subsequent introduction of a reactant, either A or B, would upset this established equilibrium and the reaction would again proceed until equilibrium conditions were re-established. Conversely, the removal of a reactant or the introduction of additional product would stimulate the back-reaction in order to reattain equilibrium. Changes in temperature and pressure may also affect equilibria by their influence upon reaction rates and concentrations.

It should be noted that an equilibrium constant can be defined for any chemical reaction. Several terms may be utilised to specify further the process under consideration. Acid–base equilibria are discussed in terms of association and dissociation constants. Stability constants are used to define ion pair formation and complexation processes. Solid solution interactions are described by means of a solubility product. Despite the differences in nomenclature, these are all equilibrium constants subject to the constraints and considerations outlined above. These are described in more detail in subsequent sections.

2.4.4 Acid–base equilibria

Acid–base equilibria play an important part in determining the composition of natural waters. Several concepts of an acid have been suggested. Arrhenius considered an acid to be a substance which upon dissolution yields a solution containing an excess of hydrogen ions (protons). Alternatively, the dissolution of a base results in an excess of hydroxyl ions. While it is convenient to represent these species as H^+ and OH^-, respectively, it must be remembered that these ions in solutions are actually associated with a varying number of water molecules, i.e. $H \cdot (H_2O)_X^+$ and $OH \cdot (H_2O)_X^-$, where $X = 1$ to 4. Under the Brønsted–Lowry concept of acids and bases, an acid was defined as a substance that could donate a proton to another substance (proton donor), while conversely a base was defined as a substance which could accept a proton from another substance. Thus, acid–base interactions were considered to involve the transfer of protons. The Lewis theory of acids and bases emphasises the role of electron pairs in acid–base reactions. A Lewis acid is a substance capable of accepting and sharing a pair of electrons, and a Lewis base acts as an electron pair donor. The Lewis and Brønsted definitions are similar for substances classified as bases because any substance capable of donating an electron pair can donate that electron pair to (i.e. accept) a proton. On the other hand, Lewis acids include substances such as oxides and metal ions as well as Brønsted acids (proton donors).

Prior to a discussion on acid–base equilibria of environmental importance, the conventions of pH must be introduced. Sorensen originally proposed a definition based on the hydrogen ion concentration, whereby

$$pH = -\log_{10}[H^+] \tag{2.134}$$

The recent trend has shifted toward the adoption of a definition utilising the hydrogen ion activity:

$$pH = -\log_{10}\{H^+\} \qquad (2.135)$$

where

$$-\log_{10}\{H^+\} = -\log_{10}[H^+] - \log_{10}\gamma_{H^+} \qquad (2.136)$$

The hydrogen ion activity is measured by means of a glass electrode which has been standardised against buffer solutions of known hydrogen ion activity. A *buffer* solution consists of the solution of a weak acid and a salt of that weak acid which will maintain a constant pH despite the addition of small amounts of an acid or base. Buffers will be further discussed later in this section. Two reference states are commonly used for the purpose of standardising electrodes. Most chemical applications utilise buffer solutions which have been prepared in pure water, thereby defining an *infinite dilution activity scale*. In this instance, $\gamma_{H^+} \to 1$ as the concentration of hydrogen ions and all other constituents approaches zero. Alternatively, the preparation of the buffer solution in a solution of fixed electrolytic composition, such as synthetic sea water, defines a *constant ionic medium activity scale*. For these systems, $\gamma_{H^+} \to 1$ as the hydrogen ion concentration approaches zero while all other components are maintained at a constant concentration. While both these approaches are equally well justified by thermodynamic considerations, the scales do differ. Accordingly, the absolute value for the magnitude of a dissociation constant (defined subsequently) will depend not only upon temperature and pressure, but also the pH scale that has been adopted. Unless otherwise indicated, the infinite dilution activity scale will be utilised here.

The simplest system to consider initially is that of pure water. The self-ionisation (dissociation) of water occurs as follows:

$$H_2O + H_2O \rightleftharpoons H_3O^+ + OH^- \qquad (2.137)$$

for which the equilibrium (dissociation) constant is defined as

$$K = \frac{\{H_3O^+\}\{OH^-\}}{\{H_2O\}^2} \qquad (2.138)$$

As discussed previously, the hydration of protons is generally assumed and so $\{H_3O^+\}$ is equivalent to $\{H^+\}$. The activity of water is generally taken to be unity. This holds true for pure water and dilute solution but will decrease as the electrolyte concentration increases. For sea water, $\{H_2O\}$ is approximately 0.98. Ignoring the activity term for water, the dissociation of water is then defined as

$$K_w = \{H^+\}\{OH^-\} \qquad (2.139)$$

where K_w is known as the *ion product of water*. It should be noted that equilibrium constants are often expressed in a fashion analogous to that used for pH. Hence, for the general case,

$$pK = -\log_{10} K \qquad (2.140)$$

Similarly

$$pK_w = -\log_{10} K_w \qquad (2.141)$$

A solution is termed *neutral* when the hydrogen ion activity equals the hydroxyl ion activity. At 24°C and 1 atmosphere pressure, $pK_w = 14.0000$ and therefore the neutral conditions for pure water correspond to pH = 7.00. As indicated in Table 2.5, pK_w is a function of temperature, thereby affecting the pH of neutrality.

The strength of an acid (or base) is determined by the extent to which a proton may be donated (or accepted). The tendency for proton transfer depends upon the ability of the acid concerned to release a proton and the facility with which the proton is accepted by a base. Hence, strengths of acid are measured relative to a common base which is generally the solvent and most often is water.

The transfer of a proton from some acid, HA, to water is denoted by the reaction

$$HA + H_2O \rightleftharpoons H_3O^+ + A^- \qquad (2.142)$$

Table 2.5. *Variations in the ion product of water with temperature*

Temperature (°C)	pK_w
0	14.9435
5	14.7338
10	14.5346
15	14.3436
20	14.1669
24	14.0000
25	13.9965
30	13.8330
35	13.6801
40	13.5348
45	13.3960
50	13.2617

for which the equilibrium (acid dissociation) constant, K_A, may be defined as:

$$K_A = \frac{\{H_3O^+\}\{A^-\}}{\{HA\}\{H_2O\}} \qquad (2.143)$$

As noted earlier, $\{H_3O^+\}$ is equivalent to $\{H^+\}$.

Similarly, the ionisation of a base, B, in water involves the transfer of a proton from the water to the base:

$$B + H_2O \rightleftharpoons BH^+ + OH^- \qquad (2.144)$$

for which the equilibrium (base dissociation) constant, K_B, is defined as

$$K_B = \frac{\{BH^+\}\{OH^-\}}{\{B\}\{H_2O\}} \qquad (2.145)$$

If we consider the specific case for the base ammonia (NH_3), the ionisation in water is represented as

$$NH_3 + H_2O \rightleftharpoons NH_4^+ + OH^- \qquad (2.146)$$

The base dissociation constant is therefore

$$K_B = \frac{\{NH_4^+\}\{OH^-\}}{\{NH_3\}} \qquad (2.147)$$

The ammonium ion, NH_4^+, can behave as an acid:

$$NH_4^+ + H_2O \rightleftharpoons NH_3 + H_3O^+ \qquad (2.148)$$

for which the acid dissociation constant is

$$K_A = \frac{\{NH_3\}\{H^+\}}{\{NH_4^+\}} \qquad (2.149)$$

From these expressions for K_A and K_B for NH_3, we see that

$$K_A K_B = K_w \qquad (2.150)$$

Also, it is apparent that water may act as a base or acid. Such behaviour is termed *amphoteric*.

Worked example
Calculate the pH of the following solutions: (i) $\{H^+\} = 3.25 \times 10^{-4}$, (ii) 0.10 M HCl, (iii) 1.0 M HCl, (iv) 0.5 M KOH and (v) 0.100 M CH_3COOH.

(i) $\{H^+\} = 3.25 \times 10^{-4}$

$$pH = -\log_{10} \{H^+\}$$
$$= -\log_{10} (3.25 \times 10^{-4})$$
$$= 3.488$$

(ii) 0.10 M HCl

HCl is a strong acid and so completely dissociates

$$HCl \rightarrow H^+ + Cl^-$$
$$[H^+] = [HCl]$$
$$= 0.10 \text{ M}$$

assuming that $\gamma_{H^+} = 1.00$, we have

$$\{H^+\} = [H^+]$$
$$= 0.10 \text{ M}$$
$$pH = 1.00$$

(iii) 1.0 M HCl

As above

$$\{H^+\} = 1.0 \text{ M}$$
$$pH = 0.00$$

(iv) 0.05 M KOH

KOH is a strong base and so completely dissociates:

$$[OH^-] = [KOH]$$
$$= 0.05 \text{ M}$$

Assuming $\gamma_{OH^-} = 1.00$, we have

$$\{OH^-\} = 0.05$$
$$pOH = -\log_{10} \{OH^-\}$$
$$= -\log_{10} (0.05)$$
$$= 1.3$$

As

$$K_w = \{H^+\}\{OH^-\}$$
$$pH = pK_w - pOH$$
$$= 14.0 - 1.3$$
$$= 12.7$$

(v) 0.100 M CH₃COOH

Acetic acid is a weak acid, $pK_A = 4.75$, and so not all of the acid will be dissociated.

$$CH_3COOH \rightarrow CH_3COO^- + H^+$$

$$K = \frac{\{CH_3COO^-\}\{H^+\}}{\{CH_3COOH\}}$$

Assuming that X moles of the acid dissociate, we have

$$[CH_3COOH\} = 0.100 - X$$
$$\{CH_3COO^-\} = X$$
$$\{H^+\} = X$$

giving

$$1.78 \times 10^{-5} = \frac{(X)(X)}{(0.100 - X)}$$

$$X^2 = 1.78 \times 10^{-5} \times 0.100 + 1.78 \times 10^{-5} \times (-X)$$

$$X^2 + 1.78 \times 10^{-5}X - 1.78 \times 10^{-6} = 0$$

This quadratic equation is solved using

$$X = \frac{-b + \sqrt{(b^2 - 4ac)}}{2a}$$

where

$$a = 1$$
$$b = 1.78 \times 10^{-5}$$
$$c = -1.78 \times 10^{-6}$$

So

$$X = \frac{1.78 \times 10^{-5} + \sqrt{[(1.78 \times 10^{-5})^2 + 4 \times 1 \times 1.78 \times 10^{-6}]}}{2 \times 1}$$

$$= 1.34 \times 10^{-3}$$

as

$$\{H^+\} = X$$
$$pH = -\log_{10}(1.34 \times 10^{-3})$$
$$= 2.873$$

Alternatively, we can make the assumption that little dissociation of CH_3COOH occurs and so the concentration does not change (i.e. $\{CH_3COOH\} = 0.100$ rather than $0.100 - X$). The equation for K is simplified to

$$1.78 \times 10^{-5} = \frac{X^2}{0.100}$$

$$X = \sqrt{(1.78 \times 10^{-6})}$$

$$= 1.33 \times 10^{-3}$$

$$pH = 2.875$$

Note that this simplification introduces less than 1% error in the final result.

The weak acid, $0.100\,M$ CH_3COOH, gives a solution of pH 2.873 while the strong acid, $0.10\,M$ HCl, gives a much more acidic solution of pH 1.0.

The dissociation constants (given as pK_A and pK_B) for acids and bases of environmental importance are presented in Table 2.6. Several acids have more than one ionisable proton and an acid dissociation

Table 2.6. *Equilibrium constants (in pure water at 25°C) for acids and bases of environmental importance*

Acid	pK_A	Base	pK_B
HCl	−3	Cl^-	17
H_2SO_4	−3	HSO_4^-	17
HNO_3	−1	NO_3^-	15
H_3O^+	0	H_2O	14
HSO_4^-	1.9	SO_4^{2-}	12.1
H_3PO_4	2.1	$H_2PO_4^-$	11.9
CH_3COOH	4.75	CH_3COO^-	9.25
H_2CO_3	6.37	HCO_3^-	7.63
H_2S	7.1	HS^-	6.9
$H_2PO_4^-$	7.2	HPO_4^{2-}	6.8
H_3BO_4	9.3	$B(OH)_4^-$	4.7
NH_4^+	9.3	NH_3	4.7
HCO_3^-	10.25	CO_3^{2-}	3.75
HS^-	14	S^{2-}	0
H_2O	14	OH^-	0

constant can be defined for each species. For phosphoric acid, H_3PO_4, three dissociation constants are given as follows:

(i) $H_3PO_4 \rightleftharpoons H^+ + H_2PO_4^-$

$$K_1 = \frac{\{H^+\}\{H_2PO_4^-\}}{\{H_3PO_4\}}$$

(ii) $H_2PO_4^- \rightleftharpoons H^+ + HPO_4^{2-}$

$$K_2 = \frac{\{H^+\}\{HPO_4^{2-}\}}{\{H_2PO_4^-\}}$$

(iii) $HPO_4^{2-} \rightleftharpoons H^+ + PO_4^{3-}$

$$K_3 = \frac{\{H^+\}\{PO_4^{3-}\}}{\{HPO_4^{2-}\}}$$

The different species of phosphoric acid and their associated pK values can be represented as

$$H_3PO_4 \underset{}{\overset{pK_1=2.1}{\rightleftharpoons}} H_2PO_4^- \underset{}{\overset{pK_2=7.2}{\rightleftharpoons}} HPO_4^{2-} \underset{}{\overset{pK_3=12.3}{\rightleftharpoons}} PO_4^{3-}$$

The predominant species at any value of pH may be readily recognised. The expression for the first dissociation constant may be given as

$$pK_1 = pH + \log \frac{\{H_3PO_4\}}{\{H_2PO_4^-\}} \tag{2.151}$$

Three different conditions may be recognised:

(A) $pH = pK_1$ $\{H_3PO_4\} = \{H_2PO_4^{2-}\}$,
(B) $pH < pK_1$ $\{H_3PO_4\} > \{H_2PO_4^{2-}\}$,
(C) $pH > pK_1$ $\{H_3PO_4\} < \{H_2PO_4^{2-}\}$.

Thus, the predominant species present when the pH is less than 2.1 will be H_3PO_4. By such reasoning, $H_2PO_4^-$ will predominate in the pH range 2.1 to 7.2; HPO_4^{2-} in the pH range 7.2 to 12.3; and PO_4^{3-} for pH > 12.3. This procedure identifies the species which exists in greatest concentration for any value of pH, however, this will not be the only species present. In sea water (salinity = 35‰, pH = 8, temperature = 20°C) the relative contribution of the species $H_3PO_4:H_2PO_4^-:HPO_4^{2-}:PO_4^{3-}$ is approximately 0:1:87:12. This concept of several coexisting species is further exemplified in Section 2.4.7.

A solution containing a mixture of a weak acid (or base) and a salt of that weak acid (or base) is known as a buffer solution. A weak acid has a low dissociation constant and so a relatively large amount of the undissociated acid (HA) would be present. Similarly, the presence of the salt ensures a relatively high concentration of the anion (A^-). Such buffer solutions are able to maintain a relatively constant pH upon the addition of small amounts of acid or base. Considering the following generalised reaction:

$$HA \rightleftharpoons H^+ + A^- \tag{2.152}$$

The addition of an acid drives this reaction to the left with the formation of HA. As noted above, both A^- and HA are in abundant supply in comparison to the amount of acid added and so their concentrations alter relatively little. Alternatively, any added base (OH^-) would react with the undissociated acid forming an anion and H_2O. Again the concentrations of HA and A^- would exhibit little overall change. The pH for the equilibrium for reaction (2.152) can be defined (i.e. re-arrange equation (2.143)) as:

$$pH_i = pK_A + \log \frac{\{A_i^-\}}{\{HA_i\}} \tag{2.153}$$

where the subscript i refers to the initial conditions. Consider again the case when a small amount of base is added to a system. From equilibrium considerations we note that $\{A_i^-\}$ will increase by an amount equal to the hydroxyl ion addition (i.e., $\{\Delta OH^-\}$) while $\{HA\}$ will decrease by that same amount. The final pH is given by

$$pH_f = K_A + \log \frac{(\{A_i^-\} + \{\Delta OH^-\})}{(\{HA_i\} - \{\Delta OH^-\})} \tag{2.154}$$

$$= pK_A + \log \frac{\{A_f^-\}}{\{HA_f\}}$$

Thus, only the ratio $\{A^-\} : \{HA\}$ is affected by adding base. As these are both present at high concentration, the pH of the solution changes only very slightly provided that $\{\Delta OH^-\}$ remains small in relation to $\{A^-\}$ and $\{HA\}$. Similar arguments apply for the addition of acid although the log term in equation (2.154) is somewhat altered. The maximum buffer effect will occur when the initial pH approximates the pK_A.

The following calculations clarify these concepts. What would be the effect of adding 0.005 moles of NaOH to a $1\,dm^3$ buffer solution containing $0.100\,M$ CH_3COOH (acetic acid) and $0.100\,M$ $Na^+CH_3COO^-$ (sodium acetate)? Because acetic acid is a weak acid

($pK_A = 4.75$, Table 2.6), assume that all the acid remains undissociated. The initial pH as given by equation (2.153) is

$$pH_i = pK_A + \log \frac{\{CH_3COO^-\}}{\{CH_3COOH\}}$$

$$= 4.75 + \log \frac{\{0.100\}}{\{0.100\}}$$

$$= 4.75 + 0$$

$$= 4.75$$

Assuming that there is no volume change, the final pH after adding the 0.005 moles of NaOH would be calculated from equation (2.154) to be

$$pH_f = pK_A + \log \frac{(0.100 + 0.005)}{(0.100 - 0.005)}$$

$$= 4.75 + \log 1.11$$

$$= 4.75 + 0.04$$

$$= 4.79$$

The overall change in pH is only 0.04 units. Compare this to the case where 0.005 moles of NaOH are added to water having an initial pH of 7. The NaOH is a strong base and so completely dissociates. Therefore,

$$\{OH^-\} = 0.005 \text{ mol } l^{-1}$$

$$pOH^- = 2.3$$

$$pH = 14.0 - 2.3 = 11.7$$

The pH change with no buffer present would be 4.7 units.

Worked example
The pH of human blood is buffered at 7.40 due to phosphoric acid species. Assuming all activity coefficients equal unity, what would be the concentration ratio of $H_2PO_4^- : HPO_4^{2-}$ at this pH?

$$H_2PO_4^- + H_2O \rightleftharpoons H_3O^+ + HPO_4^{2-}$$

$$K_2 = \frac{[H_3O^+][HPO_4^{2-}]}{[H_2PO_4^-]}$$

therefore

$$\frac{[H_2PO_4^-]}{[HPO_4^{2-}]} = \frac{[H_3O^+]}{K_2}$$

As the $pH = 7.40$, therefore $[H_3O^+] = 10^{-7.40}$. Similarly, pK_2 for phosphoric acid is 7.20 (Table 2.6) and so $K = 10^{-7.20}$

$$\frac{[H_2PO_4^-]}{[HPO_4^{2-}]} = \frac{10^{-7.40}}{10^{-7.20}} = 10^{-0.20}$$

$$= 0.63$$

Sea water is buffered at a pH near 8 by virtue of the presence of carbonic acid (H_2CO_3) and calcium carbonate ($CaCO_3$). The carbon dioxide/carbonate system presents the most important acid–base equilibria in the environment. This system is also a rather complicated one involving a gas, which not only dissolves in water but reacts with it, several aqueous species, and solid carbonate material which may be present as pure phases or solid solutions. The equilibria to be considered are:

(i) Exchange of carbon dioxide across the air–sea interface:

$$CO_2(g) \rightleftharpoons CO_2(aq) \tag{2.155}$$

(ii) Hydration

$$CO_2 + H_2O \rightleftharpoons H_2CO_3 \tag{2.156}$$

If the hydration reaction proceeds at a greater rate than the dissolution step, dissolved carbon dioxide concentrations are very low and the processes above are combined to give:

$$CO_2(g) + H_2O \rightleftharpoons H_2CO_3 \tag{2.157}$$

for which the equilibrium constant can be written:

$$K_{CO_2} = \frac{\{H_2CO_3\}}{p_{CO_2}\{H_2O\}} \tag{2.158}$$

where p_{CO2} is the partial pressure of the gaseous carbon dioxide.

(iii) Dissociation:

$$H_2CO_3 \rightleftharpoons H^+ + HCO_3^- \tag{2.159}$$

$$HCO_3^- \rightleftharpoons H^+ + CO_3^{2-} \tag{2.160}$$

for which the respective first and second dissociation constants are:

$$K_1 = \frac{\{H^+\}\{HCO_3^-\}}{\{H_2CO_3\}} \qquad (2.161)$$

$$K_2 = \frac{\{H^+\}\{CO_3^{2-}\}}{\{HCO_3^-\}} \qquad (2.162)$$

The hydrogen ion activity can be determined with a pH meter. Carbonate and bicarbonate ion activities cannot be measured. However, the concentrations may be determined (e.g. by titration as outlined below). Accordingly, one often uses apparent rather than true equilibrium constants. The former are distinguished by a prime notation. Hence, the apparent dissociation constants are

$$K_1' = \frac{\{H^+\}[HCO_3^-]}{[H_2CO_3]} \qquad (2.163)$$

$$K_2' = \frac{\{H^+\}[CO_3^{2-}]}{[HCO_3^-]} \qquad (2.164)$$

Numerical values for the constants pK_{CO_2}, pK_1' and pK_2' at different temperatures are given in Table 2.7. These values are based on a constant ionic medium scale, namely sea water with a chlorinity of 19‰ (see Section 5.3.1 for further details).

The total concentration of inorganic carbon, ΣCO_2, in solutions is given by

Table 2.7. *Equilibrium constants* for the carbonate system*

T (°C)	pK_{CO_2}	pK_1'	pK_2'
0	1.19	6.15	9.40
5	1.27	6.11	9.34
10	1.34	6.08	9.28
15	1.41	6.05	9.23
20	1.47	6.02	9.17
25	1.53	6.00	9.10
30	1.58	5.98	9.02

* Equilibrium constants given here are based on the constant ionic medium scale with a reference state of sea water having a chlorinity of 19‰ (see Section 5.3.1 for further details); adapted from Stumm & Morgan (1981).

$$\Sigma CO_2 = [H_2CO_3] + [HCO_3^-] + [CO_3^{2-}] \qquad (2.165)$$

To facilitate the discussion of pH calculations, the term alkalinity must be introduced. *Alkalinity* is a measure of the proton deficiency of a solution, not to be confused with basicity which is manifested by a high pH. The alkalinity is operationally defined by the titration of the solution to the H_2CO_3/HCO_3^- end point. This measurement may therefore be further specified as the *titration alkalinity*, TA, and corresponds to the summation of the weak acid (i.e. weaker acids than H_2CO_3) anion equivalents. For typical, well-oxygenated sea water the most important species are HCO_3^-, CO_3^{2-}, $B(OH)_4^-$ (borate) and the dissociation products of water. Hence for sea water we have

$$TA = [HCO_3^-] + 2[CO_3^{2-}] + [B(OH)_4^-] + [OH^-] - [H^+] \qquad (2.166)$$

The hydrogen and hydroxyl ion concentrations are negligible relative to the other terms and can be numerically ignored. The borate concentration is relatively small, about 3%, and can be corrected for a given sea water salinity (i.e. salt content, see Section 5.3.1) to yield the *carbonate alkalinity*, CA:

$$CA = [HCO_3^-] + 2[CO_3^{2-}] \qquad (2.167)$$

Typically, sea water carbonate alkalinities are about 2.40×10^{-3} eq dm^{-3}.* The atmospheric carbon dioxide partial pressure is 3.30×10^{-4} atm. This provides sufficient information, with the constants in Table 2.7, to calculate the pH of sea water. The following examples assume that γ_{H_2O} and $\gamma_{H_2CO_3}$ are equal to unity:

$$\{H_2CO_3\} = K_{CO_2} \times p_{CO_2}$$

(from equation 2.158)

$$\{H_2CO_3\} = 10^{-1.41} \times 3.3 \times 10^{-4}$$
$$= 1.28 \times 10^{-5}$$
$$CA = [HCO_3^-] + 2[CO_3^{2-}]$$
$$= \frac{K_1'[H_2CO_3]}{\{H^+\}} + \frac{2K_1'K_2'[H_2CO_3]}{\{H^+\}^2}$$

---

* In this context the unit eq refers to gram equivalents of hydrogen ions, defined by the fact that 1 mole of strong monobasic acid (i.e. an acid comprising one displaceable hydrogen ion per molecule) contains one gram equivalent of displaceable hydrogen. For H^+, gram equivalents are equal to gram ions.

Therefore

$$CA\{H^+\}^2 - K_1'[H_2CO_3]\{H^+\} - 2K_1'K_2'[H_2CO_3] = 0$$

This quadratic equation is solved using the following formula:

$$\{H^+\} = \frac{-b + \sqrt{(b^2 - 4ac)}}{2a}$$

where

$$a = 2.4 \times 10^{-3}$$
$$b = -10^{-6.05} \times 1.28 \times 10^{-5}$$
$$= -1.14 \times 10^{-11}$$
$$c = -2 \times 10^{-6.05} \times 10^{-9.23} \times 1.28 \times 10^{-5}$$
$$= -1.34 \times 10^{-20}$$

So

$$\{H^+\} = 5.72 \times 10^{-9}$$
$$pH = 8.24$$

The final components in the inorganic carbon system to consider are the carbonate minerals. The buffering of sea water is due to the reaction:

$$CaCO_3 + CO_2 + H_2O \rightleftharpoons Ca^{2+} + 2HCO_3^- \qquad (2.168)$$

While the preceding discussion assumes a chemical control on the pH, it should be noted that organisms may influence the pH largely by altering the *in situ* partial pressure of CO_2. Surface waters generally have a pH in the range 6 to 9. The upper limits are observed in the surface waters of lakes during summer months or when sea water is trapped in tidal rock pools. In such instances, intense photosynthetic activity decreases the carbon dioxide concentration. By Le Chatelier's principle, the equilibrium for reaction (2.159) is shifted to the left causing the pH to rise. The lower pH values are attained in fresh waters (rivers, lakes) having a low buffering capacity (i.e. low alkalinity) when CO_2 is the only acidic constituent in appreciable concentrations. For instance, the pH of pure water in equilibrium with CO_2 at atmospheric partial pressure is about 5.6. The pH of soil solutions may be as low as 4 due to microbial activity. This occurs firstly due to the oxidation of organic material which increases the CO_2 partial pressure and hence

decreases the pH. Secondly, organic acids may be released into solution. More extreme cases of low pH values require an additional acidic
input. Hence, acid rain and acid fog may have a pH below 3 due to the
dissolution of the oxidation products of SO_2 and NO_2 (see Section 5.1).
Crater lakes on volcanoes may have a pH as low as 1 due to the high
concentrations of HCl.

2.4.5 Formation of ion pairs and complexes

Electrolytic or ionic salts ($NaCl$, $CaCO_3$, $FePO_4$) upon dissolution in water dissociate into their constituent cations (Na^+, Ca^{2+},
Fe^{3+}) and anions (Cl^-, CO_3^{2-}, PO_4^{3-}). These species all exist in
hydrated forms analogous to those discussed for H^+ and OH^-. That is,
they will possess a coordinated water envelope. Oppositely charged
species will exert an electrostatic attraction. The resultant collisions
cause the transient formation of an *ion pair* in which a cation and anion,
each retaining an associated water envelope, will remain in close
contact for a short but discrete time period. The collision frequency of
oppositely charged species increases with the salt concentration. The
formation of ion pairs is enhanced in solvents with a low dielectric
constant and for species with small ionic radii.

Discrete ion pairs exist only as transient species. Nonetheless, a
dynamic equilibrium can be attained for which an equilibrium constant,
termed a *stability constant*, can be defined. For the formation of the ion
pair $NaSO_4^-$,

$$Na^+ + SO_4^{2-} \rightleftharpoons NaSO_4^- \qquad (2.169)$$

the equilibrium constant is given as

$$K = \frac{\{NaSO_4^-\}}{\{Na^+\}\{SO_4^{2-}\}} \qquad (2.170)$$

In this instance, $\log_{10} K = 0.72$. Generally, the stability of ion pairs
increases as the charge on the cation or anion increases. For $MgSO_4$,
$\log_{10} K = 2.36$.

A *complex* is formed when a central metal atom, acting as a Lewis
acid, shares a pair of electrons donated by another constituent acting as
a Lewis base. The constituent which donates the electron pair is known
as a *ligand* and may be neutral (i.e. H_2O, NH_3) or anionic (Cl^-, Br^-,
CO_3^{2-}). In contrast to an ion pair, complexes exhibit bonding that is
covalent in nature rather than electrostatic and a dehydration reaction is

involved such that the metal ion and ligand (or ligands) share a water envelope. A ligand which donates a single pair of electrons is termed *unidentate*. Ligands having more than one pair of electrons which can be donated to a central metal atom are known as *multidentate*, although the terms bidentate (two) and tridentate (three) may occasionally be used. When a multidentate ligand bonds with a central metal atom in such a way that a ring structure is formed, the complex is known specifically as a *chelate* (see Figure 2.4). Chelates exhibit an enhanced stability over related complexes involving unidentate ligands which is known as the chelate effect. Finally, the coordination number specifies the number of nearest neighbours (i.e. ligand atoms) to the central metal atom. The coordination number is indicative of the structure. Most metal ions exhibit coordination numbers of 2, 4 or 6 giving rise, respectively, to linear, tetrahedral or octahedral complexes.

The formation of complexes may be considered an equilibrium process for which an equilibrium constant, known as a stability or formation constant, can be defined. Considering the general case and ignoring charges, complex formation involving a metal (M) and some ligand (L) can be formulated as:

Figure 2.4. The structure of Zn–EDTA chelate (bond lengths not to scale) clearly showing the formation of five rings incorporating the central metal atom.

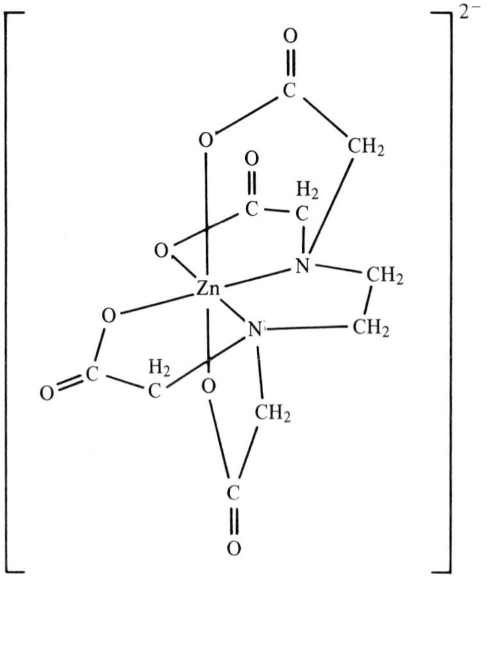

$$M + L \rightleftharpoons ML \tag{2.171}$$

whereby

$$K_1 = \frac{\{ML\}}{\{M\}\{L\}} \tag{2.172}$$

Complex formation proceeds in a stepwise manner such that

$$ML + L \rightleftharpoons ML_2 \tag{2.173}$$

$$K_2 = \frac{\{ML_2\}}{\{ML\}\{L\}} \tag{2.174}$$

From the above it can be seen that

$$\{ML\} = K_1\{M\}\{L\} \tag{2.175}$$

therefore

$$K_2 = \frac{\{ML_2\}}{K_1\{M\}\{L\}^2} \tag{2.176}$$

Combining the constants, $K_1 K_2$, we can introduce a *stability constant*, β_2, defined exclusively in terms of the metal ion activity, ligand activity and complex under consideration:

$$\beta_2 = \frac{\{ML_2\}}{\{M\}\{L\}^2} \tag{2.177}$$

Further stepwise complexation with the ligands may be feasible, leading to the general case:

$$\beta_n = \frac{\{ML_n\}}{\{M\}\{L\}^n} \tag{2.178}$$

For a given constant concentration of some metal in solution, the free ion activity of that metal will be determined by the free ion activities of all available ligands and the stability constants for all possible complexes. All metal ions can potentially form ligand bonds but the stability constants for univalent cations are so low as to make them insignificant. Several species can act as potential ligands. Inorganic ligands may be charged (F^-, Cl^-, CO_3^{2-}) or neutral (NH_3, H_2O). Important organic ligands can involve donor atoms of oxygen (carboxylate and hydroxyl groups), nitrogen (amines) and sulphur (thiols). Several factors contribute to determining stability constants (i.e. metal ion and ligand type, charge effects, steric hindrance, etc.) and a few important features of complexes can be introduced here. Table 2.8 presents the stability constants for a range of metals with the tridentate ligand nitrilotriacetate (NTA)* at 20°C in sea water. The trend in values reveals that transition metals form more stable complexes than do group II cations. The high stability constant exhibited by copper indicates that it would be very strongly complexed by NTA in sea water. As a result, the activity of the free metal ion would be greatly reduced in a closed system. This could still hold true for an open system, but, more importantly, the concentration of copper in solution would increase.

2.4.6 Solubility product

A further equilibrium to consider arises from the situation when a solid salt is in contact with a solution containing its constituent ions. Consider the case for barite, $BaSO_4$:

$$BaSO_4 \rightleftharpoons Ba^{2+} + SO_4^{2-} \tag{2.179}$$

At equilibrium (i.e. saturation)

Table 2.8. *Stability constants for various metals with nitrilotriacetate (NTA) at 20°C in sea water*

Metal	log K
Ca^{2+}	6.4
Cd^{2+}	9.5
Co^{2+}	10.6
Cu^{2+}	13.0
Fe^{2+}	10.9
Ni^{2+}	11.3
Pb^{2+}	11.8
Zn^{2+}	10.4

* NTA, chemically $N(CH_2COO)_3^{3-}$, is a synthetic complexing agent used as a water softener in some detergent products.

$$K = \frac{\{Ba^{2+}\}\{SO_4^{2-}\}}{\{BaSO_4\}} \tag{2.180}$$

and the *solubility product*,

$$K_{sp} = \{Ba^{2+}\}\{SO_4^{2-}\}$$

The instantaneous value of the ion product ($\{Ba^{2+}\}\{SO_4^{2-}\}$), whether at saturation or otherwise, is termed the *ion activity product, IAP*. Three relationships between K_{sp} and *IAP* can be considered:

(i) $IAP < K_{sp}$, the solution is undersaturated and the salt will dissolve.

(ii) $IAP = K_{sp}$, equilibrium exists between the salt and the saturated solution.

(iii) $IAP > K_{sp}$ the solution is supersaturated and the salt will precipitate until $IAP = K_{sp}$.

The solubility (i.e. total concentration) of a constituent in solution may be determined by the solubility of the least soluble salt. This will be true for simple chemical systems but can be applied to environmental circumstances only with considerable caution. The equilibrium under consideration must be free from biological or kinetic hindrance. Ion pairing and complexation in solution must be taken into account together with the possible effects of solid solution of the precipitate (i.e. activity of the solid would be less than unity).

For barite in sea water:

$$K_{sp} = 1.08 \times 10^{-10} \, mol^2 \, l^{-2}$$

at 25°C, and

$$IAP = 5.2 \times 10^{-11} \, mol^2 \, l^{-2}$$

Thus, sea water is apparently under-saturated with respect to barite. Any pure barium sulphate introduced in the sea would dissolve. However, natural barite contains a significant amount of celestite, $SrSO_4$. The barium and strontium sulphates exist in solid solution. Consequently, the activity of the solid phase is less than unity which in turn reduces the equilibrium barium ion activity. It has been shown that barium in sea water is in equilibrium with natural barite, $(Ba_{0.66}Sr_{0.33})SO_4$.

Worked example

The solubility product of anhydrite ($CaSO_4$) is $4.2 \times 10^{-5} \, mol^2 \, l^{-2}$ and that of fluorite (CaF_2) is 3.98×10^{-11} mol^{3}l^{-3}. What is the solubility for each compound in water?

(i) $CaSO_4$

$$CaSO_4 \rightarrow Ca^{2+} + SO_4^{2-}$$

Thus, the saturated solution contains equal amounts of Ca^{2+} and SO_4^{2-}:

$$[Ca^{2+}] = [SO_4^{2-}] = X$$

where X is solubility in mol dm^{-3}.

$$K_{sp} = X^2$$
$$X = \sqrt{(4.2 \times 10^{-5})}$$
$$= 6.48 \times 10^{-3} \text{ mol dm}^{-3}$$

The solubility of anhydrite is 6.48×10^{-3} mol dm^{-3}.

(ii) CaF_2
From the equation

$$CaF_2 \rightarrow Ca^{2+} + 2F^-$$

the concentration of F^- is twice that of Ca^{2+}. The solubility S is equal to the concentration of Ca^{2+}, hence

$$K_{sp} = [Ca^{2+}][F^-]^2$$
$$= (S)(2S)^2$$
$$= 4S^3$$

$$S = \sqrt[3]{\left(\frac{K_{sp}}{4}\right)}$$

$$= \sqrt[3]{\left(\frac{3.98 \times 10^{-11}}{4}\right)}$$

$$= 2.15 \times 10^{-4} \text{ mol dm}^{-3}$$

The solubility of fluorite is 2.15×10^{-4} mol dm^{-3}.

2.4.7 Equilibrium diagrams

As discussed in the previous sections, chemical equilibrium data can be used to establish the speciation of a constituent in solution or the equilibrium concentration of a saturated solution in contact with an undissolved salt. Graphical representation of equilibrium data facilitates the assessment of the relative importance of a large number of species under a prescribed set of conditions and the shift in equilibrium that follows a perturbation of those conditions. Due to the environmental importance and variability of pH, this is often used as a *master variable* in the construction of an equilibrium diagram. As outlined below, several formats are available for the graphical presentation of equilibria data.

In a *distribution diagram*, the ratio or percentage contribution of the activity (or concentration) of a species to the total activity of that constituent is plotted as a function of some master variable. The construction of a distribution diagram is illustrated using the example of the speciation of phosphoric acid as a function of pH. Four species must be considered (Section 2.4.4):

$$\{P\}_T = \{H_3PO_4\} + \{H_2PO_4^-\} + \{HPO_4^{2-}\} + \{PO_4^{3-}\} \quad (2.181)$$

where $\{P\}_T$ is the total activity for all the phosphate species. The activities for each of the dissociated species can be defined in terms of the activity of phosphoric acid using dissociation constants. Therefore,

$$\{H_2PO_4^-\} = \frac{K_1\{H_3PO_4\}}{\{H^+\}} \quad (2.182)$$

$$\{HPO_4^{2-}\} = \frac{K_2\{H_2PO_4^-\}}{\{H^+\}} = \frac{K_1K_2\{H_3PO_4\}}{\{H^+\}^2} \quad (2.183)$$

$$\{PO_4^{3-}\} = \frac{K_3\{HPO_4\}^{2-}}{\{H^+\}} = \frac{K_1K_2K_3\{H_3PO_4\}}{\{H^+\}^3} \quad (2.184)$$

and

$$\{P\}_T = \{H_3PO_4\} \left(1 + \frac{K_1}{\{H^+\}} + \frac{K_1K_2}{\{H^+\}^2} + \frac{K_1K_2K_3}{\{H^+\}^3}\right) \quad (2.185)$$

As a shorthand notation, define D as

$$D = \left(1 + \frac{K_1}{\{H^+\}} + \frac{K_1K_2}{\{H^+\}^2} + \frac{K_1K_2K_3}{\{H^+\}^3}\right)^{-1} \quad (2.186)$$

The distribution diagram can then be constructed using the following four expressions:

$$\frac{\{H_3PO_4\}}{\{P\}_T} = D \tag{2.187}$$

$$\frac{\{H_2PO_4^-\}}{\{P\}_T} = \frac{K_1D}{\{H^+\}} \tag{2.188}$$

$$\frac{\{HPO_4^{2-}\}}{\{P\}_T} = \frac{K_1K_2D}{\{H^+\}^2} \tag{2.189}$$

$$\frac{\{PO_4^{3-}\}}{\{P\}_T} = \frac{K_1K_2K_3D}{\{H^+\}^3} \tag{2.190}$$

The speciation of phosphoric acid as a function of pH is illustrated in Figure 2.5. The distribution diagram is independent of total phosphate activity. A vertical line drawn at any pH gives the relative contribution of each species. Hence, the area of predominance for any species is clearly indicated. These regions are bounded by the inflexion points in the distribution curves. The pH for each inflexion point corresponds to the pK of the acid.

Worked example
What percentage of each species of H_3PO_4 exists at pH 8.00?

Figure 2.5. A distribution diagram showing the percentage composition of phosphoric acid species as a function of pH.

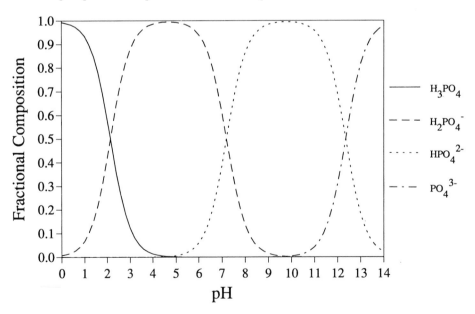

$$\text{pH} = 8.00, \qquad \{H^+\} = 10^{-8.00}$$
$$K_1 = 10^{-2.1}, \qquad K_2 = 10^{-7.2}, \qquad K_3 = 10^{-12.3}$$

$$D = \left(1 + \frac{K_1}{\{H^+\}} + \frac{K_1 K_2}{\{H^+\}^2} + \frac{K_1 K_2 K_3}{\{H^+\}^3}\right)^{-1}$$

$$= \left(1 + \frac{10^{-2.1}}{10^{-8.00}} + \frac{10^{-2.1} \times 10^{-7.2}}{10^{-16.00}} + \frac{10^{-2.1} \times 10^{-7.2} \times 10^{-12.3}}{10^{-24.00}}\right)^{-1}$$

$$= (1 + 10^{5.9} + 10^{6.7} + 10^{2.4})^{-1}$$

$$= (1 + 7.94 \times 10^5 + 5.0 \times 10^6 + 2.51 \times 10^2)^{-1}$$

$$= (5.81 \times 10^6)^{-1}$$

$$= 1.72 \times 10^7$$

Note that $\{P\}_T = 100\%$

(i) $\{H_3PO_4\}$

$$\{H_3PO_4\} = \{P\}_T \times D$$
$$= 100 \times 1.72 \times 10^{-7}$$
$$= 1.72 \times 10^{-5}\%$$

(ii) $\{H_2PO_4^-\}$

$$\{H_2PO_4^-\} = \{P\}_T \times D \times \frac{K_1}{\{H^+\}}$$
$$= 100 \times 1.72 \times 10^{-7} \times 10^{5.9}$$
$$= 13.7\%$$

(iii) $\{HPO_4\}^{2-}$

$$\{HPO_4^{2-}\} = \{P\}_T \times D \times \frac{K_1 K_2}{\{H^+\}^2}$$
$$= 100 \times 1.72 \times 10^{-7} \times 10^{6.7}$$
$$= 86.2\%$$

(iv) $\{PO_4^{3-}\}$

$$\{PO_4^{3-}\} = \{P\}_T \times D \times \frac{K_1 K_2 K_3}{\{H^+\}^3}$$
$$= 100 \times 1.72 \times 10^{-7} \times 10^{2.4}$$
$$= 4.32 \times 10^{-3}\%$$

Thus, at pH 8 the two predominant species are HPO_4^{2-} (86.2%) and $H_2PO_4^-$ (13.7%). The percentage composition for each species as a function of pH is depicted in Figure 2.5.

Master variables other than pH may be employed. Figure 2.6 depicts the speciation of mercury as a function of chloride ion concentration. This distribution diagram is constructed in a fashion analogous to that for the speciation of phosphoric acid and the salient features are outlined below:

$$[Hg]_T = [Hg^{2+}] + [HgCl^+] + [HgCl_2] + [HgCl_3^-] + [HgCl_4^{2-}] \quad (2.191)$$

$$D = (1 + K_1[Cl^-] + \beta_2[Cl^-]^2 + \beta_3[Cl^-]^3 + \beta_4[Cl^-]^4)^{-1} \quad (2.192)$$

$$\frac{[Hg^{2+}]}{[Hg]_T} = D \quad (2.193)$$

$$\frac{[HgCl^-]}{[Hg]_T} = DK_1[Cl^-] \quad (2.194)$$

$$\frac{[HgCl_2]}{[Hg]_T} = D\beta_2[Cl^-]^2 \quad (2.195)$$

$$\frac{[HgCl_3^-]}{[Hg]_T} = D\beta_3[Cl^-]^3 \quad (2.196)$$

Figure 2.6. The percentage composition of mercury chlorocomplexes as a function of pCl^- (i.e. $pCl^- = -\log[Cl^-]$).

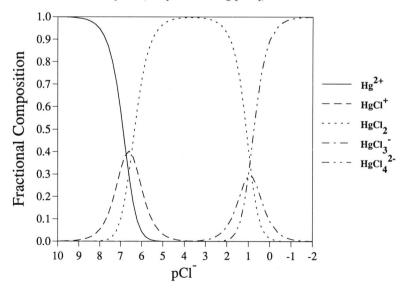

$$\frac{[HgCl_4^{2-}]}{[Hg]_T} = D\beta_4[Cl^-]^4 \qquad (2.197)$$

The constants used in this example are: $\log K_1 = 6.74$, $\log K_2 = 6.48$, $\log K_3 = 0.85$ and $\log K_4 = 1.00$. At the chloride ion concentration of sea water, $[Cl^-] = 0.559 \, mol \, l^{-1}$, the major species are $HgCl_3^-$ and $HgCl_4^{2-}$. Of course, this treatment ignores complexation with other halides, hydroxyl groups and organic ligands.

Equilibrium data can also be graphically presented in the form of a *logarithmic diagram*. In such cases, the activity (or concentration) of each species is plotted versus a master variable in the form of a log–log plot. In the case where activities are utilised throughout, this type of representation may be specifically referred to as an *activity–activity diagram*. The logarithmic diagram for the phosphoric acid system is given in Figure 2.7 as it is very similar to the distribution diagram given earlier. In contrast with distribution diagrams, an additional constraint must be imposed for the construction of a logarithmic diagram in that the total activity (or concentration) of the constituent must be stipulated. The speciation of phosphoric acid in Figure 2.7 assumes a total concentration condition of $\log[P]_T = -1.00$. Equations (2.187) to (2.190) can be arranged to give the concentration of any of the species for a given pH. Note that the slopes of the lines will be 0, ± 1 or ± 2, except in the immediate vicinity of the cross-over points where two species will exist in equal concentration. A further example of an activity–activity diagram of importance for examining environmental systems is the pH–pE diagram. These are dealt with in Section 2.5.5.

Figure 2.7. A logarithmic concentration diagram for phosphoric acid species with the condition $\log[P]_T = -1.00$. c_i refers to the concentration of each species, i.

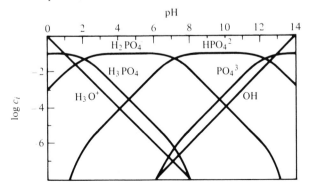

2.5   Electrochemistry

2.5.1 Introduction

Electrochemistry is the study of chemical reactions which involve the transfer of electrons from one reactant to another. Two processes occur simultaneously. *Reduction* occurs when a reactant gains electrons and goes to a more negative oxidation state. Ignoring charges, this is exemplified by the general case:

$$A_{ox} + ne^- \rightarrow A_{red} \qquad (2.198)$$

where A_{ox} and A_{red} refer to oxidised and reduced forms of element A. Conversely, during oxidation, a reactant loses electrons and goes to a more positive oxidation state. The general case can be represented as:

$$B_{red} \rightarrow B_{ox} + ne^- \qquad (2.199)$$

where again B_{ox} and B_{red} refer to oxidised and reduced forms of element B. Each general expression above is termed a *half-reaction*. This is in recognition of the fact that neither process can occur independently. For instance, the electrons are transient in solution and so an equilibrium constant for a half-reaction cannot be meaningfully defined. Reduction and oxidation proceed concurrently, and two half-reactions combine to give a *redox couple*. For the general case, the redox couple is given as:

Reduction half-reaction: $A_{ox} + ne^- \rightarrow A_{red}$

Oxidation half-reaction: $\qquad B_{red} \rightarrow B_{ox} + ne^-$

Redox couple: $\qquad A_{ox} + B_{red} \rightarrow A_{red} + B_{ox} \qquad (2.200)$

Thus, a redox couple involves the reaction of a reductant (B_{red}) with an oxidant (A_{ox}). The *reductant* or reducing agent is the reactant which itself loses electrons and so is oxidised. The *oxidant* or oxidising agent gains electrons and so is reduced.

2.5.2 Standard electrode potential

All half-reactions may be expressed in terms of reduction or oxidation. For example, the Fe^{2+}/Fe^0 half-reactions are

reduction: $\qquad Fe^{2+} + 2e^- \rightarrow Fe^0 \qquad (2.201)$

oxidation: $\qquad Fe^0 \rightarrow Fe^{2+} + 2e^- \qquad (2.202)$

For discussion here, an *electrode* can consist of a conducting metal in contact with a solution of its ions. For example,

$$Fe^0(s)/Fe^{2+} \qquad\qquad (2.203)$$

for which the possible half-reactions are given above. Alternatively, an electrode may be an inert metal such as platinum in contact with a solution containing a redox couple:

$$Pt^0(s)/Fe^{3+}, Fe^{2+} \qquad\qquad (2.204)$$

Different substances vary in their tendency to undergo reduction or oxidation.

The *potential* is a measure of the reactant's ability to be reduced or oxidised. The potential of a single half-reaction (i.e. a single electrode) cannot be measured. Instead the potential difference between two half-reactions (i.e. two electrodes) is determined. Potentials are measured relative to the *standard hydrogen electrode* (SHE) also known as the *normal hydrogen electrode* (NHE). This consists of a platinised platinum wire in a solution of unit hydrogen ion activity over which a hydrogen gas at 1 atmosphere pressure is bubbled. This is represented as:

$$Pt^0(s)/H_2(f = 1 \text{ atm, gas})), H^+ (a = 1, \text{aqueous}) \qquad (2.205)$$

The half-reaction which occurs:

$$2H^+ + 2e^- \rightarrow H_2(g) \qquad\qquad (2.206)$$

is arbitrarily assigned a potential of 0.000 V. The SHE is impracticable for use in natural waters, and consequently other reference electrodes are used.

For comparative purposes all half-reactions are written as reductions. The potential difference between the SHE and any reduction half-reaction (for which all ions in solution exist at unit activity) is called the *standard electrode potential*, $E^\ominus$. This may also be termed the standard reduction potential because of the convention universally adopted of writing half-reactions as reduction processes. Table 2.9 lists the standard electrode potentials for a range of half-reactions in order of their reduction affinity. The following sign convention is adopted:

(i) A positive $E^\ominus$ indicates that the oxidised form is a better oxidising agent than H^+.

(ii) A negative $E^\ominus$ indicates that the oxidised form is a poorer oxidising agent than H^+.

Thus, a constituent with a high $+E^{\ominus}$ will be a strong oxidant (e.g. O_2) while a constituent with a low $-E^{\ominus}$ will be a strong reductant (e.g. Fe). In combining half-reactions, an oxidant will react with a reductant whose $E^{\ominus}$ is lower. This means that the half-reaction with the highest

Table 2.9. *Standard electrode potentials at 25°C*

Half-reaction	$E^{\ominus}$ (V)	pE
$F_2 + 2H^+ + 2e^- \rightarrow 2HF$	3.06	51.78
$Ce(IV) + e^- \rightarrow Ce^{3+}$ (in 1 M $HClO_4$)	1.70*	28.76
$Ce(IV) + e^- \rightarrow Ce^{3+}$ (in 1 M HNO_3)	1.61*	27.24
$MnO_4^- + 8H^+ + 5e^- \rightarrow Mn^{2+} + 4H_2O$	1.51	25.55
$Ce(IV) + e^- \rightarrow Ce^{3+}$ (in 1 M H_2SO_4)	1.44*	24.37
$Cr_2O_7^{2-} + 14H^+ + 6e^- \rightarrow 2Cr^{3+} + 7H_2O$	1.33	22.50
$Ce(IV) + e^- \rightarrow Ce^{3+}$ (in 1 M HCl)	1.28*	21.66
$O_2 + 4H^+ + 4e^- \rightarrow 2H_2O$	1.23	20.81
$MnO_2 + 4H^+ + 2e^- \rightarrow Mn^{2+} + 2H_2O$	1.21	20.47
$HNO_2 + H^+ + e^- \rightarrow NO + H_2O$	1.00	16.92
$2Hg^{2+} + 2e^- \rightarrow Hg_2^{2+}$	0.905	15.31
$Ag^+ + e^- \rightarrow Ag$	0.800	13.54
$Hg_2^{2+} + 2e^- \rightarrow 2Hg$	0.789	13.35
$Fe^{3+} + e^- \rightarrow Fe^{2+}$	0.771	13.05
$O_2 + 2H^+ + 2e^- \rightarrow H_2O_2$	0.682	11.54
$I_2 + 2e^- \rightarrow 2I^-$	0.535	9.05
$Cu^+ + e^- \rightarrow Cu$	0.521	8.82
$H_2SO_3 + 4H^+ + 4e^- \rightarrow S + 3H_2O$	0.45	7.6
$Cu^{2+} + 2e^- \rightarrow Cu$	0.337	5.70
$Cu^{2+} + e^- \rightarrow Cu^+$	0.153	2.59
$Sn^{4+} + 2e^- \rightarrow Sn^{2+}$	0.15	2.5
$S + 2H^+ + 2e^- \rightarrow H_2S$	0.141	2.39
$2H^+ + 2e^- \rightarrow H_2$	0.000	0.00
$Pb^{2+} + 2e^- \rightarrow Pb$	-0.126	-2.13
$Sn^{2+} + 2e^- \rightarrow Sn$	-0.136	-2.30
$Ni^{2+} + 2e^- \rightarrow Ni$	-0.250	-4.23
$Cd^{2+} + 2e^- \rightarrow Cd$	-0.403	-6.82
$Fe^{2+} + 2e^- \rightarrow Fe$	-0.440	-7.45
$Zn^{2+} + 2e^- \rightarrow Zn$	-0.763	-12.91
$Mn^{2+} + 2e^- \rightarrow Mn$	-1.18	-19.97
$Ca^{2+} + 2e^- \rightarrow Ca$	-2.87	-48.56
$Li^+ + e^- \rightarrow Li$	-3.04	-51.44

* These are more properly formal potentials, $E^{\ominus\prime}$, for the solution conditions stated.

$E^{\ominus}$ is written as a reduction, and the other half-reaction is reversed and written as an oxidation (for which the standard electrode potential must also be reversed). Then the standard electrode potential for the redox couple can be calculated. This is exemplified below using the O_2/Fe system. The half-reactions are:

$$O_2 + 4H^+ + 4e^- \rightarrow 2H_2O \qquad E^{\ominus} = 1.229 \text{ V} \qquad (2.207)$$

$$Fe^{2+} + 2e^- \rightarrow Fe \qquad E^{\ominus} = -0.440 \text{ V} \qquad (2.208)$$

Oxygen will be reduced and elemental iron will be oxidised. Hence, this half-reaction is written in the oxidation form and multiplied by a factor of two in order to balance the electrons for the coupled process, noting that the $E^{\ominus}$ is independent of reaction stoichiometry. Therefore we have:

$$O_2 + 4H^+ + 4e^- \rightarrow 2H_2O \qquad\qquad E^{\ominus} = 1.229 \text{ V}$$

$$2Fe \rightarrow 2Fe^{2+} + 4e^- \qquad\qquad E^{\ominus} = 0.440 \text{ V}$$

$$\overline{}$$

$$O_2 + 4H^+ + 2Fe \rightarrow 2H_2O + 2Fe^{2+} \qquad E^{\ominus} = 1.669 \text{ V} \quad (2.209)$$

Note that the $E^{\ominus}$ for Fe^{3+}/Fe^{2+} (0.771 V) is less than that for O_2 (1.229 V) and so the reaction process given above would be an intermediate step only in the oxidation of Fe to Fe^{3+} (i.e. rust formation).

Values given for the standard electrode potentials in Table 2.9 refer to all species existing in solution at unit activity. In some instances that metal ion may undergo complexation with anions. This reduces the metal ion activity. As the composition of these complexes may not be known, a solution containing unit activity of the metal ion cannot be prepared. To overcome this difficulty, a *formal electrode potential*, $E^{\ominus\prime}$, is defined to be the standard electrode potential under specified solution conditions for a redox couple with the oxidised and reduced forms at 1 M concentration. This accounts for the various $E^{\ominus}$ values for Ce(IV) in different acidic media.

Worked example

If Sn^{2+} is added to a solution of Fe^{3+}, what reaction occurs and what would be the resulting $E^{\ominus}$ for the reaction?

$$Fe^{3+} + e^- \rightarrow Fe^{2+} \qquad E^{\ominus} = 0.771 \text{ V}$$

$$Sn^{4+} + 2e^- \rightarrow Sn^{2+} \qquad E^{\ominus} = 0.15 \text{ V}$$

In this redox couple, the Fe^{3+} has the highest $E^{\ominus}$ and so must act as the oxidant. To balance electrons, this half-reaction must be doubled. Therefore,

$$2Fe^{3+} + 2e^- \rightarrow 2Fe^{2+} \qquad E^{\ominus} = 0.771 \text{ V}$$

The Sn^{2+} is oxidised and so is written as oxidation half-reaction, and the sign of the $E^{\ominus}$ is reversed, giving

$$Sn^{2+} \rightarrow Sn^{4+} + 2e^- \qquad E^{\ominus} = -0.15 \text{ V}$$

The overall reaction would be

$$2Fe^{3+} + Sn^{2+} \rightarrow 2Fe^{2+} + Sn^{4+}$$
$$E^{\ominus} = 0.771 \text{ V} - 0.15 \text{ V}$$
$$= 0.62 \text{ V}$$

Note that for the reduction of Fe^{2+}:

$$Fe^{2+} + 2e^- \rightarrow Fe$$

the $E^{\ominus}$ is -0.440 V. This is lower than the $E^{\ominus}$ for Sn^{4+}/Sn^{2+} and so Sn^{2+} cannot reduce Fe^{2+} to Fe.

2.5.3 Electrochemical cells

There are two classifications of electrochemical cells, galvanic and electrolytic. A *galvanic* (or voltaic) cell is a combination of two half-reactions in such a way that spontaneous chemical reactions at the electrodes produce a current. Usable electrical energy is generated; a lead acid battery is an example of a galvanic cell. An *electrolytic* cell is a galvanic cell in which a potential is applied across the electrodes to oppose any current which might otherwise flow. Electrical energy is consumed rather than produced. The electrode processes are non-spontaneous and opposite to those anticipated in the galvanic cell itself.

Regardless of the cell type, the electrodes are designated as either cathodes or anodes depending upon the electrode process that occurs rather than the charge that is exhibited. A *cathode* is an electrode at which reduction occurs. In a galvanic cell the cathode is positively charged and attracts anions in solution. In an electrolytic cell the cathode is negatively charged and cations are reduced at the electrode surface. An *anode* is an electrode at which oxidation occurs. It is negatively charged in galvanic cells and cations are attracted. Alternatively, in electrolytic cells the anode is positively charged and anions are oxidised.

2.5.4 Gibbs energy relationships and the concept of pE

The Gibbs energy change is related to the electrode potential by

$$\Delta G = -nFE \qquad (2.210)$$

where n is the number of electrons involved in the stoichiometric reaction and F is Faraday's constant. *Faraday's constant* is the quantity of electrical energy associated with one equivalent (i.e. one gram electron) of chemical change at an electrode (i.e. the transfer of 6.02×10^{23} electrons), namely 9.648×10^4 C. It should be noted here that several environmental and geochemical texts use the notation Eh for potential rather than E. There is no difference in meaning, the h indicates that the redox scale is referenced to the hydrogen electrode as discussed earlier. We know from discussions in Section 2.3.9 that a large but negative value for a Gibbs energy change favours a reaction process. Using equation (2.210) to compare electrochemical reactions, the more favoured reaction thermodynamically is that involving the largest positive value for E.

At standard state conditions, the standard Gibbs energy change is:

$$\Delta G^{\ominus} = -nFE^{\ominus} \qquad (2.211)$$

Substitution from equation (2.130) gives the relationship between the equilibrium constant and the standard electrode potential

$$-nFE^{\ominus} = -\text{RT} \ln K \qquad (2.212)$$

Rearrangement and substitution for numerical constants gives

$$\log K = \frac{nE^{\ominus}}{0.0591} \qquad (2.213)$$

Hence, the hypothetical equilibrium constant for the oxygen half-reaction (equation (2.207)) is

$$\log K = \frac{4 \times 1.23}{0.0591}$$

$$\log K = 83.2$$

Worked example
Determine the $\Delta G^{\ominus}$ and K for the reaction (at 25°C):

$$2Fe^{3+} + Sn^{2+} \rightarrow 2Fe^{2+} + Sn^{4+} \qquad E^{\ominus} = 0.62 \text{ V}$$

$$\Delta G^{\ominus} = -nFE^{\ominus}$$
$$= -2 \times 9.65 \times 10^4 \times 0.62$$
$$= -120 \text{ kJ}$$

$$\log_{10} K = \frac{\Delta G^{\ominus}}{2.303RT}$$

$$= \frac{-120 \times 10^3}{2.303 \times 8.314 \times 298.15}$$

$$= 21.0$$
$$K = 1.00 \times 10^{21}$$

Two different terminologies may be utilised in describing the oxidation or reduction intensity of water masses. Firstly, one can consider the redox potential E, also known as Eh as discussed previously. A high value of E indicates that oxidising conditions are prevalent. Conversely a low E signifies reducing conditions. The second, but related, system used the pE notation. Just as pH is defined as the negative log activity of protons in solution:

$$\text{pH} = -\log \{H^+\} \qquad (2.214)$$

so can the pE be defined as the negative log activity of electrons in solution:

$$pE = -\log \{e^-\} \qquad (2.215)$$

This is a hypothetical parameter, in that free electrons do not exist in solution any more than do free protons. As is the case for the redox potential, a high pE denotes oxidising conditions while a low pE indicates reducing conditions. The approximate range for natural waters is -10 to 17 (for pH 4 to 10). Note that pE is not the log of the potential, E, but related to electron activity. The relationship between pE and E is

$$pE = \frac{F}{2.303RT} E \quad \left(= \frac{E}{0.0591} \text{ at } 25°C \right) \qquad (2.216)$$

and for the standard state

$$pE^{\ominus} = \frac{F}{2.303RT} E^{\ominus} \qquad (2.217)$$

This can be related to the equilibrium constant by the following expression:

$$pE^{\ominus} = \frac{1}{n}\log_{10} K \tag{2.218}$$

In the usage of $pE^{\ominus}$ notation, and especially in the comparison of redox reactions, it is convenient to express half-reactions as one electron process. Hence, the reduction of iron(II) given earlier:

$$Fe^{2+} + 2e^{-} \rightarrow Fe \tag{2.219}$$

would be expressed in the form:

$$\tfrac{1}{2}Fe^{2+} + e^{-} \rightarrow \tfrac{1}{2}Fe \tag{2.220}$$

2.5.5 The Nernst equation

The standard potential of an electrode is the potential that is established when all constituents exist in their standard states (i.e. unit activity for all dissolved species). The potential of the electrode will therefore be different when the redox constituents are not in their standard state. The *Nernst equation*, given below, is used to calculate the electrode potential for varying activities of the redox species. For the general half-reactions:

$$A_{ox} + ne^{-} = A_{red} \tag{2.221}$$

the Nernst equation is

$$E = E^{\ominus} + \frac{RT}{nF}\ln\frac{\{A_{ox}\}}{\{A_{red}\}} \tag{2.222}$$

where E is the potential (in volts) of the electrode versus SHE, $E^{\ominus}$ is the standard electrode potential, R is the universal gas constant, T is the absolute temperature in kelvin, n is the number of electrons involved in the stoichiometric reaction, and F is Faraday's constant. At 25°C, substitution of various numerical constants yields the following equation:

$$E = E^{\ominus} + \frac{0.0591}{n}\log_{10}\frac{\{A_{ox}\}}{\{A_{red}\}} \tag{2.223}$$

In concentration units this becomes

$$E = E^{\ominus'} + \frac{0.0591}{n} \log_{10} \frac{[A_{ox}]}{[A_{red}]} \tag{2.224}$$

Alternatively, the Nernst equation can be expressed in pE notation as

$$pE = pE^{\ominus} + \frac{1}{n} \log_{10} \frac{\{A_{ox}\}}{\{A_{red}\}} \tag{2.225}$$

In the application of the Nernst equation, half-reactions involving H^+ or OH^- must include such species in the numerator if they are reactants and in the denominator if they are products. Accordingly, the Nernst equations for the half-reactions of O_2 and Fe cited earlier (equations 2.207 and 2.208) are:

$$E = 1.229 + \frac{0.0591}{4} \log_{10} \frac{\{O_2\}\{H^+\}^4}{\{H_2O\}} \tag{2.226}$$

$$E = -0.440 \text{ V} + \frac{0.0591}{2} \log_{10} \frac{\{Fe^{2+}\}}{\{Fe\}} \tag{2.227}$$

Worked example
For the half-reaction $Fe^{2+} + 2e^- \rightarrow Fe$, what is the potential for solutions of Fe^{2+} activity 1.00 and 1.00×10^{-3} at 25°C and 60°C?

$$E = E^{\ominus} + \frac{RT}{nF} \ln \frac{\{A_{ox}\}}{\{A_{red}\}}$$

For Fe^{2+}/Fe this is

$$E = -0.440 + \frac{2.303RT}{2F} \log \{Fe^{2+}\}$$

Note that $\{Fe\}$ is taken to be unity.

Calculations at 25°C:
(i) $a_{Fe^{2+}} = 1.00$
As this is the standard state condition
$$E = E^{\ominus} = -0.440 \text{ V}$$

(ii) $a_{Fe^{2+}} = 1.00 \times 10^{-3}$

$$E = -0.440 + \frac{2.303 \times 8.314 \times 298.15}{2 \times 9.649 \times 10^4} \log(1.00 \times 10^{-3})$$

$$= -0.440 - 0.089$$

$$= -0.529 \text{ V}$$

Calculations at 60°C:

$$\frac{2.303RT}{nF} = \frac{2.303 \times 8.314 \times 333.15}{2 \times 9.649 \times 10^4}$$

$$= 3.31 \times 10^{-3}$$

(i) $a_{Fe^{2+}} = 1.00$

$$E = -0.440 + 3.31 \times 10^{-3} \times \log(1.00)$$
$$= -0.440 + 0$$
$$= -0.440 \text{ V}$$

(ii) $a_{Fe^{2+}} = 1.00 \times 10^{-3}$

$$E = -0.440 + 3.31 \times 10^{-3} \times \log(1.00 \times 10^{-3})$$
$$= -0.440 - 0.099$$
$$= -0.539 \text{ V}$$

The limits of pE in water can be determined using the Nernst equation. The stability regime of water is determined by the reduction and oxidation of water. The reduction of water defines the lower limit to pE:

$$2H_2O + 2e^- \rightarrow H_2 + 2OH^- \tag{2.228}$$

Water is in equilibrium with its dissociation products, H^+ and OH^-, and so the potential for the reduction of water can be given by

$$H^+ + e^- \rightarrow \tfrac{1}{2}H_2 \tag{2.229}$$

$$pE = pE^\ominus + \log_{10} \frac{H^+}{f_{H_2}^{\frac{1}{2}}} \tag{2.230}$$

As p$E^\ominus$ is zero by definition and the boundary condition commonly used is a hydrogen fugacity, f_{H_2}, of unity, this gives:

$$pE = -pH \tag{2.231}$$

The upper boundary for the stability of water is given by the oxidation reaction:

$$\tfrac{1}{4}O_2 + H^+ + e^- \rightarrow \tfrac{1}{2}H_2O \tag{2.232}$$

$$pE = pE^\ominus + \log f_{O_2}^{\frac{1}{4}}\{H^+\} \tag{2.233}$$

The boundary condition used is an oxygen fugacity of unity. The upper pE limit becomes:

$$pE = 20.75 - pH \qquad (2.234)$$

Thus, for water with a pH of 7.00, the pE limits given by equations (2.231) and (2.234) are -7.00 to 13.75.

2.5.6 pH–pE diagrams

The construction of activity–activity diagrams was outlined in Section 2.4.7. A particularly useful activity–activity diagram is that using both pH and pE as master variables. Figure 2.8 illustrates the approximate pH–pE values for typical environmental waters. This diagram also indicates the stability regime of water. The lower boundary, limited by water reduction, is therefore defined by equation (2.231). The upper boundary is given by equation (2.234), this being the condition for water oxidation.

The pE for natural waters can be calculated using equation (2.233). The unknowns which must be measured are pH and the O_2 partial pressure (concentration). For example, the pE for a surface water in

Figure 2.8. Approximate pH–pE regimes for typical environmental waters. The boundaries for water stability are also indicated.

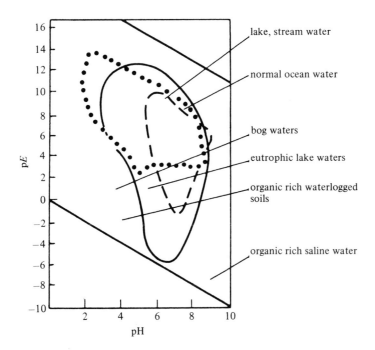

equilibrium with the atmosphere ($p_{O_2} = 0.21$ atm) and having a pH of 8 would be:

$$pE = pE^{\ominus} + \log f_{O_2}^{\frac{1}{4}}\{H^+\} \qquad (2.235)$$

$$pE = pE^{\ominus} - pH + \log f_{O_2}^{\frac{1}{4}} \qquad (2.236)$$

$$pE = 20.75 - 8 + \log (0.21)^{\frac{1}{4}} = 12.58 \qquad (2.237)$$

According to this calculation, a typical pE for normal, well-oxygenated water in equilibrium with atmospheric oxygen would be about 13.

The construction of a simple pH–pE diagram is demonstrated using the Fe–H_2O system. The only species considered are Fe^{2+}, Fe^{3+}, $Fe(OH)_2(s)$ and $Fe(OH)_3(s)$. The predominance area for each species within the stability regime of water is shown in Figure 2.9. Equilibria related to the boundaries in this figure are listed in Table 2.10. Detailed calculations for each boundary are given below for the Fe–H_2O system

Figure 2.9. A pH–pE diagram for the Fe system in water. The stability regimes for only Fe^{3+}, Fe^{2+}, $Fe(OH)_3(s)$ and $Fe(OH)_2(s)$ are determined assuming a maximum soluble Fe(II) or (III) concentration of 1.0×10^{-5} M.

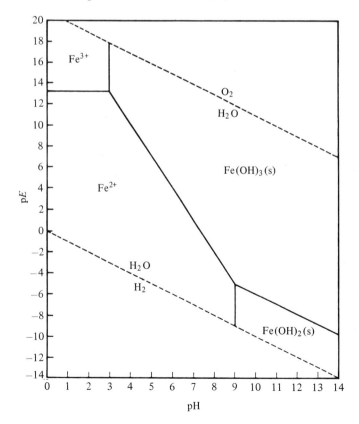

assuming a maximum Fe^{2+} or Fe^{3+} concentration in solution of 1.0×10^{-5} M, all activity coefficients equal unity and the activity of pure solids and water is unity.

The stability boundaries for water were calculated earlier. The upper and lower limits, respectively, are:

$$pE = 20.75 - pH \qquad (2.238)$$

$$pE = -pH \qquad (2.239)$$

Boundary 1
Under conditions of high pE and low pH, Fe^{3+} and Fe^{2+} may both exist in solution. These conditions are calculated from the Nernst equation:

$$pE = 13.05 + \log \frac{[Fe^{3+}]}{[Fe^{2+}]} \qquad (2.240)$$

Due to equilibrium at the boundary,

$$[Fe^{3+}] = [Fe^{2+}] \qquad (2.241)$$

Therefore, the ferrous–ferric ion boundary is independent of pH and is given by

$$pE = 13.05 \qquad (2.242)$$

Boundary 2
Free ferric ions exist only under conditions of high pE (>13.05) and low pH. From (2) in Table 2.10 we see that a decrease in pH drives the reaction to the left causing the precipitation of $Fe(OH)_3(s)$. The pH at which this occurs is defined by the solubility product of $Fe(OH)_3$:

$$K_{sp} = \frac{[Fe^{3+}]}{\{H^+\}^3} = 9.1 \times 10^3 \, l^2 \, mol^{-2} \qquad (2.243)$$

For a maximum dissolved iron concentration of 1.00×10^{-5} M we get:

Table 2.10. *Equilibria in the Fe–H₂O system*

Boundary	Equilibrium reaction	
(1)	$Fe^{3+} + e^- \rightleftharpoons Fe^{2+}$	$pE^{\ominus} = 13.05$
(2)	$Fe(OH)_3(s) + 3H^+ \rightleftharpoons Fe^{3+} + 3H_2O$	$K_{sp} = 9.1 \times 10^3 \, l^2 \, mol^{-2}$
(3)	$Fe(OH)_2(s) + 2H^+ \rightleftharpoons Fe^{2+} + 2H_2O$	$K_{sp} = 8.0 \times 10^{12} \, l \, mol^{-1}$
(4)	$Fe(OH)_3(s) + H^+ + e^- \rightleftharpoons Fe(OH)_2(s) + H_2$	
(5)	$Fe(OH)_3(s) + 3H^+ + e^- \rightleftharpoons Fe^{2+} + 3H_2O$	

$$\{H^+\} = \left(\frac{[Fe^{3+}]}{K_{sp}}\right)^{1/3} \tag{2.244}$$

$$\{H^+\} = \left(\frac{1.00 \times 10^{-5}}{9.1 \times 10^3}\right)^{1/3}$$

$$pH = 2.99$$

This boundary is independent of pE.

Boundary 3
In an analogous fashion to that outlined above, the pH boundary between free ferrous ions and ferrous hydroxide can be calculated:

$$K_{sp}^* = \frac{[Fe^{2+}]}{\{H^+\}^2} = 8.0 \times 10^{12} \tag{2.245}$$

$$\{H^+\} = \left(\frac{[Fe^{2+}]}{K_{sp}^*}\right)^{1/2} \tag{2.246}$$

$$= \left(\frac{1.00 \times 10^{-5}}{8.0 \times 10^{12}}\right)^{1/2}$$

$$pH = 8.95$$

Boundary 4
The boundary between ferric hydroxide and ferrous hydroxide is a redox boundary and so can be calculated using the Nernst equation (i.e. equation 2.240). The free ferric and ferrous ions in solution are defined by the solubilities of their respective hydroxides which can be determined from equations (2.243) and (2.245). Hence, substitution with the Nernst equation gives

$$pE = pE^\ominus + \log\frac{[Fe^{3+}]}{[Fe^{2+}]}$$

$$pE = 13.05 + \log\left(\frac{K_{sp}\{H^+\}^3}{K_{sp}^*\{H^+\}^2}\right) \tag{2.247}$$

$$pE = 13.05 + \log\frac{K_{sp}}{K_{sp}^*} - pH \tag{2.248}$$

$$pE = 4.1 - pH \tag{2.249}$$

These four boundaries delineate the stability regimes of Fe^{3+} and $Fe(OH)_2$. The boundary between Fe^{2+} and $Fe(OH)_3$ must extend from one to the other of the regimes already defined.

Boundary 5
This could in fact be simply drawn by connecting the curves of the two stability regimes already defined. However, the mathematical equation is given by the Nernst equation for which the ferric ion concentration is calculated from the solubility of ferric hydroxide. Hence,

$$pE = pE^{\ominus} + \log \frac{[Fe^{3+}]}{[Fe^{2+}]}$$

$$= 13.05 + \log \left(\frac{K_{sp}\{H^+\}^3}{[Fe^{2+}]} \right) \tag{2.250}$$

$$pE = 13.05 + \log \left(\frac{K_{sp}}{[Fe^{2+}]} \right) - 3\,pH$$

$$pE = 13.05 + \log \left(\frac{9.1 \times 10^3}{1.0 \times 10^{-5}} \right) - 3\,pH$$

$$pE = 22.0 - 3\,pH \tag{2.251}$$

Thus, the five boundaries define the four stable regimes for individual constituents in this simplified $Fe-H_2O$ system. It should be appreciated that in environmental systems several further constituents would have to be considered. This list includes oxides (i.e. FeO, Fe_2O_3, Fe_3O_4) as well as sulphides (FeS, FeS_2).

2.6   Chemistry of surfaces and colloids

2.6.1 Introduction

Surface chemistry is the study of chemical processes that occur at an interface. An *interface* is the boundary region between two phases encompassing the complete zone where properties (i.e. electrical forces and chemical composition) differ from those exhibited in either of the bulk phases. For the phases gas (g), liquid (l) and solid (s), five types of interfaces exist: g–l, g–s, l–l, l–s and s–s. A *colloid* is a particle in the size range 1 nm to 1 μm. It must be appreciated that both the upper and lower limits are arbitrarily defined and may vary by a factor of two. Colloidal systems are two component systems containing discrete particles, in the defined size range, in a circumambient dispersing medium. Note that colloidal systems are said to be stable when the discrete particles can exist for long time periods. An important feature of colloidal systems is the large surface area to volume ratio of the small particles. Surface chemistry at the interface region of a colloid system may control macroscopic physical properties of the system.

Three types of colloidal systems exist. Firstly, macromolecular material, such as proteins or humic substances, may exist in true solutions. Secondly, a colloidal dispersion consists of discrete particles (disperse phase) suspended in a dispersion medium producing in the simplest case a two phase system. Eight categories of colloidal dispersions exist for the following disperse phase–dispersion medium systems: l–g (liquid aerosol), s–g (solid aerosol), g–l (foam), l–l (emulsion), s–l (sol or hydrosol, where l is water), g–s (solid foam), l–s (solid emulsion), and s–s (solid suspension). Thirdly, association colloids form when the concentration of a *surfactant* (a molecule with a strong affinity to accumulate at an interface) exceeds a critical value and aggregation occurs. Colloids may be further described as either hydrophilic or hydrophobic. A *hydrophilic* colloid has an affinity towards water and will tend to form a stable suspension in water. Such colloids are characterised by polar functional groups ($-O^-$, $-OH_2^+$). Conversely, *hydrophobic* colloids are water-repellent, do not form stable suspensions in water, and are characterised by non-polar functional groups (e.g. long chain hydrocarbons).

Surface chemistry at the liquid–solid interface is very important in environmental chemistry. Particles in natural waters exhibit a surface charge and hence an electric double layer (EDL) is formed. The EDL determines adsorption characteristics, ion-exchange properties and aggregation processes. Such phenomena can control the behaviour and distribution of organic and inorganic trace constituents in natural waters.

Thermodynamic considerations

There is work associated with creating a surface. If you consider any condensed phase, a molecule in the interior will experience a spherically symmetric attraction to all adjacent molecules. Molecules at the interface experience a net inward attraction normal to the surface. This force is termed the *interfacial tension*, σ, (usually called *surface tension*, γ, for air–liquid interfaces). This attraction reduces the number of molecules and increases their inter-molecular distances relative to molecules in the interior. To expand the surface area, A_s, molecules from the interior would have to be moved to the surface. This would require breaking bonds and is therefore non-spontaneous. The work associated with expanding the surface area is equal to

$$dw = \sigma \, dA_s \qquad (2.252)$$

The expression for Gibbs energy (equation 2.105) can be expanded to include work involved in surface deformations. At constant T, P and n_i, the Gibbs energy change is

$$dG = V \, dP - S \, dT + \Sigma \mu_i \, dn_i + \sigma \, dA_s \qquad (2.253)$$

The surface tension of water at 20°C is 72.75 mN m^{-1}. The addition of electrolytes causes the surface tension to increase as a function of solute concentration. The surface tension of sea water (salinity $= 35‰$) at 20°C is 73.53 mN m^{-1}. The ions occur in greater concentration in the bulk phase than at the surface. As the attractive force between a solute ion and a water molecule is greater than between water molecules themselves, localised order (structuring) of the water occurs and the surface tension increases. Alternatively, addition of surfactants will cause the surface tension to decrease. Soil solutions may have surface tensions as low as 45 mN m^{-1} at 20°C due to the presence of fulvic and humic substances (Sections 3.4.5 and 5.4.3). A reduction in surface tension allows easier penetration of aqueous solutions into rocks and soils of low porosity. This can become manifest by accelerated chemical weathering.

2.6.2 Adsorption

The extent to which an interface influences the distribution and concentration of a component i in two adjacent phases (α, β) can be evaluated from its *surface concentration* or *surface excess*, Γ_i. The total number of moles of i (n_i) is distributed between the two phases $(n_{i,\alpha}$ and $n_{i,\beta})$ and the surface $(n_{i,s})$, giving

$$n_i = n_{i,s} + n_{i,\alpha} + n_{i,\beta} \qquad (2.254)$$

The number of moles in each of the two phases can be expressed in terms of their concentration (C) and volume (V). Substituting these terms and rearranging gives

$$n_{i,s} = n_i - C_{i,\alpha}V_\alpha - C_{i,\beta}V_\beta \qquad (2.255)$$

The number of moles at the interface is, by analogy to the concentration–volume terms for the bulk phase, related to the surface concentration, Γ_i, and surface area, A_s. Therefore,

$$\Gamma_i A_s = n_i - C_{i,\alpha}V_\alpha - C_{i,\beta}V_\beta \qquad (2.256)$$

This formulation, originally established by Gibbs, assumes that the interface is a plane surface of no volume.

From the above discussion, it is evident that the surface concentration is defined as

$$\Gamma_i = \frac{n_{i,s}}{A_s} \qquad (2.257)$$

Adsorption is *positive* when $\Gamma_i > 0$. In a physical sense, component i accumulates at the interface (i.e. is surface-active). This arises because the attractive force between solvent molecules is greater than that between solute and solvent molecules. Positive adsorption causes a marked decrease in the surface tension and is exemplified by hydrocarbons in water.

Conversely, *negative* adsorption occurs when $\Gamma_i < 0$ and solute molecules are excluded from the interface. Solute–solvent molecular attraction is high and the component is termed surface-inactive. Surface tension is increased slightly as exemplified by NaCl in sea water (see above). As suggested by this discussion, the surface concentration may be expressed in terms of the surface tension, γ. At constant pressure and temperature, this is given by the expression

$$\Gamma_i = \frac{d\gamma}{d\mu_i} \qquad (2.258)$$

or given in terms of the activity is known as the *Gibbs adsorption isotherm*:

$$\Gamma_i = -\frac{1}{RT}\left(\frac{d\gamma}{d\ln a_i}\right) \qquad (2.259)$$

The relationship at a given temperature between the equilibrium surface concentration (or fractional surface coverage where Γ_i cannot be determined) and solute concentration is known as an *adsorption isotherm*.

Several expressions for adsorption isotherms may be derived. The *Langmuir adsorption isotherm* has the form:

$$X/M = abC(1 + aC) \tag{2.260}$$

where X is the weight of a solute sorbed by M grams of solid, C is the equilibrium solute concentration, a and b are constants, whereby $1/a$ equals the concentration when $1/2$ of the available adsorption sites are occupied and b is the total number of available adsorption sites. The Langmuir adsorption isotherm is applicable for monolayer adsorption onto a homogeneous surface when no interaction occurs between adsorbed species. The *Freundlich adsorption isotherm* is often used to describe specific adsorption processes, and is given by

$$X/M = kC^{1/n} \quad (n > 1) \tag{2.261}$$

where k and n are constants of no theoretical significance. More complex adsorption isotherms have been developed to accommodate adsorption at heterogeneous surfaces and interactions between adsorbed species.

Non-specific adsorption

Non-specific or physical adsorption involves weak long range interactions such as electrostatic attraction or van der Waals forces. Multi-layer adsorption is possible. As the attractive forces are relatively weak, the adsorbed species retain their coordinated water molecules and hence can approach no closer than the radii of their hydration spheres (termed outer Helmholtz plane, see Section 2.6.4). While physical adsorption is not dependent upon the chemical identity of the adsorbed ion, that is not to say that adsorption is equally probable for all solute species. The electrostatic attraction of counter-ions (i.e. having a charge opposite to that of the solid substrate) is favoured for ions with a high charge and small hydration radii. For this reason, trivalent cations (Al, Fe) will be adsorbed in preference to univalent (Na, K) and divalent (Ca, Mg) cations. Other factors act in favour of the trivalent species. Al and Fe tend to form polymeric oxyhydroxide cations; their large size ensures an important contribution to adsorption due to van der Waals forces. Additionally, the adsorption of a polymeric species displaces a number of monomeric species. As this tends to increase the disorder in the solution, entropy considerations favour the adsorption of polymeric species. Ion-exchange processes are a manifestation of physical adsorption and are discussed in more detail in Section 2.6.5.

Specific adsorption

Specific adsorption or chemisorption is due to short range forces such as H-bonding and π orbital interactions. As attractive forces are greater than those observed in physical adsorption, adsorbed species lose their coordinated water molecules. The distance of closest approach, known as the inner Helmholtz plane (Section 2.6.4), becomes a function of the ionic radii. Chemisorption is limited to monolayer coverage of the substrate.

Specific adsorption of transition metals onto oxide and hydroxide surfaces, especially those of Mn, Fe, Al and Si, have been well documented. These oxyhydroxide surfaces tend to adsorb water which undergoes dissociation to give a hydroxylated surface:

$$X\!-\!O + H_2O \rightarrow -X(OH)_2 \qquad (2.262)$$

where X refers to a surface atom (Fe, Mn, etc.) associated with that element's oxide structure. Free metal ions, M, in solution may adsorb onto the surface by displacement of one hydrogen ion leading to complexation with the oxygen:

$$-X\!-\!O\!-\!H + M^{Z+} \rightarrow -X\!-\!O\!-\!M^{(Z-1)+} + H^+ \qquad (2.263)$$

This is termed chelation, where two or more hydrogen ions are eliminated as shown by:

$$
\begin{array}{l}
-X\!-\!OH \\
\quad | \qquad\quad + M^{Z+} \rightarrow \\
-X\!-\!OH
\end{array}
\qquad
\begin{array}{l}
-X\!-\!O \\
\quad | \qquad\ \ \diagdown \\
\quad | \qquad\quad M^{(Z-2)+} + 2H^+ \qquad (2.264)\\
\quad | \qquad\ \ \diagup \\
-X\!-\!O
\end{array}
$$

Not only can the free metal ion be specifically adsorbed onto the surface, but also metal complexes, ML_n^{Z+}, may be adsorbed. This may occur by substitution of one or more hydrogen ions by ML_n^{Z+}:

$$-X\!-\!O\!-\!H + ML_n^{Z+} \rightarrow -X\!-\!O\!-\!ML_n^{(Z-1)+} + H^+ \qquad (2.265)$$

or, alternatively, a metal–metal bond may form following the displacement of hydroxyl ions:

$$-X\!-\!O\!-\!H + ML_n^{Z+} \rightarrow -X\!-\!ML_n^{(Z+1)+} + OH^- \qquad (2.266)$$

It should be noted that from a thermodynamic point of view all these processes (i.e. reactions 2.263 to 2.266) can be treated as equilibria. Hence, equilibrium constants can be defined and in several instances determined experimentally. This allows adsorption processes to be accommodated in models of trace element speciation in natural waters.

The most obvious manifestation of the importance of surface chemistry in the environment is the accumulation of trace elements in manganese nodules. Manganese (or also known as ferromanganese) nodules are found in most natural waters where sedimentation rates are slow. The nodules consist of alternating layers of Mn and Fe oxides. These oxide layers have extremely high adsorption capacities for transition metals. Deep-sea nodules may contain 0.5% by weight of each Cu, Ni and Co. These would have been scavenged from circumambient sea water in which such elements exist in the 0.1 to 10 nmol kg^{-1} range. This represents an enrichment of several orders of magnitude.

2.6.3 Surface charge

There are several mechanisms by which suspended solids (particulates and colloids) may exhibit a surface charge in natural waters. The charge may be an intrinsic property of a crystalline structure. A net negative charge may be produced in solids susceptible to crystal defects (i.e. vacant cation positions) or cation substitution. Clay minerals consist of layered structures of tetrahedral SiO_4 and octahedral AlO_6. Substitution of Si(IV) by Al(III) occurs at tetrahedral sites. At octahedral sites, Al(III) may be replaced by Mg(II) or Fe(II). The differential dissolution of simple electrolytic salts may produce a surface charge. Consider the case for barite ($BaSO_4$):

$$BaSO_4(s) \rightarrow Ba^{2+}(aq) + SO_4^{2-}(aq) \qquad (2.267)$$

Dissolution of the cations and anions may occur independently. A surface charge will develop whenever the rate of dissolution of one species exceeds that of the other. Proteins and polysaccharides exhibit a negative charge in sea water. The charge on colloidal organic matter comes from the dissociation of acidic functional groups.

Some adsorption mechanisms are responsible for producing or altering the surface charge. Suspended solids in sea water tend to exhibit negative surfaces due to the specific adsorption of anionic organic compounds. Alternatively, the solid surface may react with dissolved constituents or water. Dissolved metals in natural waters, especially Fe and Mn, form oxide coatings on suspended solids. Such metal oxide coatings, the frayed edges of clay minerals (i.e. broken Al—O and Si—O bonds) and natural oxides (MgO, SiO_2) present in water may be hydrolysed. As a result, the surfaces are covered with hydroxide groups which exhibit amphoteric behaviour:

$$—XO(s)^- + H^+(aq) \rightarrow —XOH(s) \qquad (2.268)$$

$$—XOH(s) + H^+(aq) \rightarrow —XOH_2^+(s) \qquad (2.269)$$

The hydroxide behaves as a Brønsted acid or base at low and high pH, respectively. In this case, H^+ ions act as potential determining ions, but other cations may exhibit this behaviour. The *point of zero change* (PZC) is the negative log of the activity (i.e. pH for H^+) for which the surface exhibits no net charge. In this case, the surface concentration of Brønsted acid and base sites are equal:

$$[—XOH_2^+] = [—XO^-] \qquad (2.270)$$

Table 2.11 presents the PZC for some solids found in natural waters.

The absorption of transition metals onto oxide and hydroxide surfaces is influenced to a great extent by the pH of the solution. At low pH

Table 2.11. *The point of zero charge (PZC) for some mineral surfaces*

Mineral	pH_{PZC}
α–Al_2O_3	9.1
α–$Al(OH)_3$	5.0
γ–$AlOOH$	8.2
CuO	9.5
Fe_3O_4	6.5
α–$FeOOH$	7.8
γ–Fe_2O_3	6.7
'$Fe(OH)_3$' (amorph)	8.5
MgO	12.4
δ–MnO_2	2.8
β–MnO_2	7.2
SiO_2	2.0
$ZrSiO_4$	5
Feldspars	2–2.4
Kaolinite	4.6
Montmorillonite	2.5
Albite	2.0
Chrysotile	>12

Reproduced from Stumm & Morgan (1981).

little or no metal cation adsorption occurs. The high hydrogen ion concentration drives the equilibria (2.268) and (2.269) to the right leading to a positively charged surface. Conversely, at high pH there is significant cation adsorption. The pH for the transition between these two extreme cases depends upon the metal ion and the adsorption substrate considered. Figure 2.10 illustrates the adsorption of four metals onto a quartz surface.

There are several environmental consequences of this observed phenomenon. In the context of today's debate on acid rain, the decrease in pH that occurs in fresh waters would lead to desorption of transition metals. There is considerable evidence to show that it is the dissolved metal ions which constitute the bioavailable (i.e. that portion of the metal loading in a natural water that can be absorbed by organisms) and hence toxic metal level. Hence, one of the several effects of acid rain may be to increase dissolved metal concentrations by desorption.

2.6.4 Electric double layer

Solids suspended in natural waters exhibit a surface charge (usually negative). This creates an ionic atmosphere in the adjacent

Figure 2.10. Experimental and computed adsorption isotherms for Fe(III), Cr(III), Co(III) and Ca(II) onto quartz. Curves, theoretical; points, experimental data. (From James & Healy, 1972.)

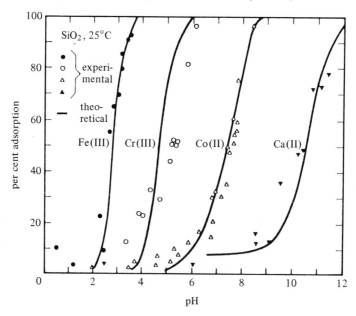

fluid affecting the concentration of electrolytes and determining the orientation of water molecules (i.e. due to the dipole moment). Thus, there is a segregation of positive and negative charges normal to the interface and considerable structuring in the fluid adjacent to the solid. The *electric double layer* (EDL) consists of the charged layer within the solid and the diffuse zone in the fluid in which the potential generated by the surface charge decays to that of the bulk solution. For electrolyte in solution, *co-ions* have the same charge as that exhibited by the suspended solid. Conversely, *counter-ions* (*gegen-ions*) are oppositely charged. We shall consider here three models to describe the structure within the fluid portion of the EDL.

In the simplest case, the fluid portion of the EDL comprises a Stern layer and a Gouy layer (Figure 2.11a). The *Stern layer* consists of a layer of ions in contact with the charged surface. In this simplest model, only non-specific adsorption occurs. Therefore, the Stern layer consists of any counter-ions electrostatically attracted to the surface. No co-ions are present and no chemical bonding occurs. The attractive forces are weak and hence the counter-ions retain their hydration spheres. Counter-ions can approach the surface no closer than the Stern plane, at a distance equivalent to the radii of the hydration sphere. The *Gouy layer* is the diffuse zone affected by the surface potential, which consequently has an electrolytic composition different to the bulk solution. The surface potential in the Gouy layer decays due to enhanced concentrations of counter-ions relative to co-ions. Furthermore, counter-ions and co-ions are more or less, respectively, abundant in comparison with the bulk solution (i.e. where the potential approaches zero). The thickness of the Gouy layer depends upon the ionic strength of the solution.

A model of the EDL which includes specific adsorption retains the Gouy layer as discussed above but requires modification of the Stern layer (Figure 2.11b). Specific adsorption involves chemical bonding of counter-ions to the charged surface of the solid rather than simply an electrostatic attraction. Because of the enhanced binding energy, counter-ions may approach closer to the surface. Such counter-ions will not have water of hydration. The locus of centres of these specifically adsorbed counter-ions is known as the *inner Helmholtz plane* (IHP) while the *outer Helmholtz plane* (OHP) is the locus of centres of the hydrated counter-ions which are electrostatically attracted to the surface.

A third situation to consider occurs when specific adsorption pro-

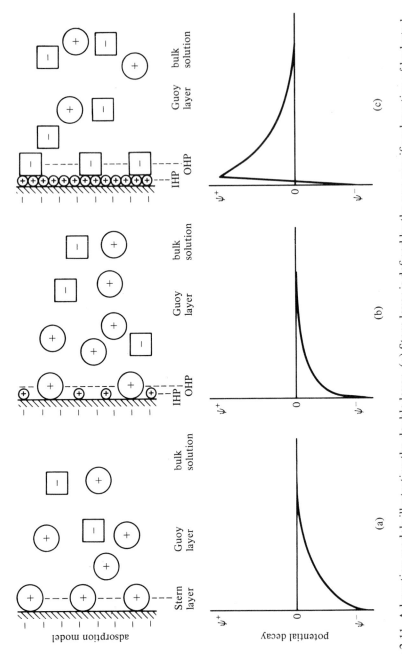

Figure 2.11. Adsorption models illustrating the double layer. (a) Stern layer is defined by the non-specific adsorption of hydrated counter-ions. The Gouy layer is the diffuse zone where the potential is less than that exhibited in the bulk solution. (b) The Stern layer is subdivided into two parts. The inner Helmholtz plane (IHP) is defined by specific adsorption of free counter-ions, while the outer Helmholtz plane (OHP) is defined by non-specific adsorption of hydrated counter-ions. (c) Surface charge reversal can occur when super-equivalent chemisorption of counter-ions occurs.

ceeds to such an extent that *charge reversal* ensues (Figure 2.11c). In this case, the potential exerted by counter-ions within the IHP is greater than the original surface charge. In a natural system this would be exemplified by a suspended solid with an anionic charge specifically adsorbing cations. When more cations than were necessary to balance the anionic charge had been adsorbed, the suspended particle would then behave as a positively charged particle. Subsequent specific or non-specific (electrostatic) adsorption of anions would occur. The potential decay in the Gouy layer would result from a relative abundance of anions rather than cations.

2.6.5 Ion-exchange processes

Mineral particles (i.e. clay minerals, zeolites, freshly precipitated oxyhydroxides) and particulate organic material can sorb cations from natural waters with the concurrent release of an equivalent amount of other cations into solution. This process is termed *cation-exchange*. The cation-exchange capacity (CEC) of a component is the summation of exchangeable cations (including hydrogen ions). This is expressed as milli-equivalents per 100 grams (meq/100 g) of material and usually measured by the uptake of ammonium ions from 1 M ammonium acetate solution at pH 7 although other cations may be utilised. Table 2.12 lists the specific surface area and the cation-exchange capacity of several materials of environmental significance.

Table 2.12. *The specific surface area and cation-exchange capacities of several sorption active materials*

Material	Specific surface area (m^2/g)	Cation-exchange capacity (meq/100 g)
Calcite (<2 μm)	12.5	—
Kaolinite	10–50	3–15
Illite	30–80	10–40
Chlorite	—	20–50
Montmorillonite	50–150	80–120
Freshly precipitated $Fe(OH)_3$	300	10–25
Amorphous silicic acid	—	11–34
Humic acids from soils	1900	170–590

From Förstner & Wittmann (1981).

Values for CEC are pH-dependent and vary as a function of the ions occupying the exchange sites.

There are several factors which will enhance the affinity of a cation, M^+, for an exchange site:

(i) *Concentration in solution.* As $[M^+]$ increases there is an increase in the fractional surface coverage Γ_{M^+}.

(ii) *Oxidation state.* An increase in the oxidation state of an element favours its accumulation at the surface; the order of affinity being

$$M^+ < M^{2+} < M^{3+}$$

(iii) *Charge density of the hydrated cation.* The greater the charge density (i.e. a decrease in the diameter of the hydrated cation), the greater its affinity for an exchange site. The order for groups I and II metal cations in increasing charge density is

$$Ba < Sr < Ca < Mg < Cs < Rb < K < Na < Li$$

The riverine transport of material in equilibrium with river water to the sea will cause the cation-exchange of Ca^{2+} and K^+ by Mg^{2+} and Na^+ in the marine environment. Ion-exchange processes are of considerable importance in determining soil solution composition and will be considered further in Section 5.4.

2.6.6 Aggregation of colloids and particles

Aggregation occurs when a colloidal system becomes destabilised. Constituents in natural waters are partitioned between dissolved and particulate components. These will have different residence times because particles experience gravitational effects. While sorption processes allow the interaction between dissolved and particulate fractions, aggregation generally initiates sedimentation. Hence, adsorption followed by aggregation and settling acts as a possible pathway from solution to the sediments.

The aggregation of colloids is usually prevented by electrostatic repulsion. Two classifications of aggregation mechanism are recognised. These are dependent upon the process by which the repulsion is overcome. *Coagulation* involves decreasing the repulsion between colloid particles by some inorganic means such that aggregation between similar components may proceed. *Flocculation* depends upon the interaction of organic material associated with the colloidal particles. These act as bridging compounds and essentially enmesh the colloids in an organic matrix. Often, however, these terms are treated as inter-changeable in common usage.

Coagulation can be promoted by increasing the ionic strength of the solution. This may be termed double-layer compression. The increased concentration of ions in solution causes enhanced surface concentrations of counter-ions, thereby decreasing the surface charge. The higher ionic strength of the solution also ensures that the surface potential decreases in a shorter distance than that of similar net charge in a dilute solution. The combined effect diminishes the electrostatic repulsion between particles. The colloidal system is therefore destabilised and aggregation occurs. Such a mechanism may contribute to the sedimentation of riverborne suspended particles in an estuary; however, this would be an oversimplification of estuarine processes. Coagulation is utilised in water treatment in order to purify domestic supplies. Aluminium or iron(III) salts are added to the water and these elements form colloidal hydroxides which in turn scavenge heavy metals from solution. Electrostatic repulsion between colloids is decreased by adjusting the pH to approximately the PZC of the hydroxide. Aggregation and sedimentation ensues.

During flocculation, the electrostatic repulsion between colloids is minimised by the adsorption of organic surfactants. Bonding between the organic compounds occurs causing bridging between colloids. The resulting large particles, called flocs, attain sufficient size to settle. In natural waters, humic substances can act as important agents in promoting flocculation. This is especially true in estuarine environments where iron is deposited as hydroxide–humate flocs. In fresh waters the humic coating on colloidal particles may influence the surface properties so as to stabilise the colloid.

Questions

(1) Sulphur dioxide is removed from smokestack gases by reaction with calcium carbonate forming anhydrite ($CaSO_4$). Write the reaction equation and calculate $\Delta H_f^{\ominus}$ and $\Delta S^{\ominus}$ for the reaction. Calculate the $\Delta G^{\ominus}$ for this reaction at 25°C and 150°C.

(2) Determine the $\Delta G^{\ominus}$ and K for the reaction

$$2NO(g) + O_2(g) \rightarrow 2NO_2(g)$$

(3) Determine the ionic strength of sea water with an approximate composition of 0.48 M NaCl, 0.04 M $MgCl_2$, 0.01 M $MgSO_4$, 0.01 M $CaSO_4$ and 0.005 M K_2SO_4.

(4) Drinking water may be dosed with 1.00 mg dm^{-3} F^-. What is the maximum concentration of Ca^{2+}, in g dm^{-3}, which can be kept in solution assuming that the solubility is determined by fluorite? (K_{sp} of $CaF_2 = 3.98 \times 10^{-11}$.)

(5) What is the concentration of $HgCl_3^-$ and $HgCl_4^{2-}$ in sea water given that the total concentration of Hg^{2+} is 2 ng dm^{-3} and $[Cl^-]$ is 0.559 mol dm^{-3}? Assume all activity coefficients equal unity.

(6) What is the concentration of each species of H_2CO_3 for a solution at pH 7.8 containing 1 mmol dm^{-3} of H_2CO_3 at 10°C? Is this solution undersaturated or supersaturated with respect to atmospheric CO_2? ($pCO_2 = 3.40 \times 10^{-4}$ atm.)

(7) Calculate the potential of the following electrodes at 25°C:
- (i) 0.05 M $SnCl_2$ in contact with Sn;
- (ii) 0.5 M $SnCl_2$ in contact with Sn;
- (iii) 1.0 M $SnCl_2$ in contact with Sn.

(8) An iron analysis of a sample containing haematite (Fe_2O_3) is carried out as follows:
- (i)   sample dissolved in HCl;
- (ii)   iron reduced with Sn^{2+};
- (iii)   excess Sn^{2+} reduced with Hg^{2+};
- (iv) Fe^{2+} titrated with K_2CrO_7.

Write balanced reactions for the above and determine $E^{\ominus}$ for each redox couple.

(9) Calculate the partial pressure of O_2 in equilibrium with the following typical waters (assuming $\gamma_{O_2} = 1$):
- (i)   normal ocean water: pH = 7.8, pE = 10.0;
- (ii)   eutrophic lake water: pH = 5.5, pE = 1.0;
- (iii)   organic-rich waterlogged soil: pH = 4.0, pE = −2.0.

(10) What iron species predominates in each of the waters in question (9)?

(11) What are the order and molecularity of the following elementary reactions?

$$CO + OH \rightarrow CO_2 + H$$
$$NO + O_3 \rightarrow NO_2 + O_2$$
$$O_3 \rightarrow O_2 + O$$

(12) Show how the reaction

$$2NO + O_2 \rightarrow 2NO_2$$

could proceed via two steps, each of second order, but the overall reaction show third order kinetics.

(13) Write down the Arrhenius equation and describe the significance of each term. Why do most reactions proceed more rapidly as temperature is increased?

(14) In an atmosphere containing initially 30 ppb of NO_2 calculate the extent of conversion to nitric acid by reaction with the hydroxyl radical after five hours if the rate constant is $2.5 \times 10^2 \, \text{ppm}^{-1} \, \text{s}^{-1}$ and the concentration is $10^6 \, \text{cm}^{-3}$. Assume: (i) no replenishment of NO_2; (ii) a constant NO_2 concentration. Express your results in ppb of HNO_3.

References

Alberty, R. & Daniels, F. (1980). *Physical Chemistry*. New York: John Wiley & Sons.

Dasent, W. (1988). *Inorganic Energetics: an introduction*, 2nd edn. Cambridge University Press.

Förstner, U. & Wittmann, G. (1981). *Metal Pollution in the Aquatic Environment*, p. 486. Heidelberg: Springer-Verlag.

James, R.O. & Healy, T. W. (1972). *J. Colloid Interface Sci.* **40**, 42–52.

Stumm, W. & Morgan, J. (1981). *Aquatic Chemistry*, 2nd edn. New York: John Wiley & Sons.

Sverdrup, H.U., Johnson, M.W. & Fleming, R. H. (1942). *The Oceans*, p. 1087. Englewood Cliffs, NJ: Prentice Hall.

Further reading

Alberty, R.A. (1987). *Physical Chemistry*, 7th edn. New York: John Wiley & Sons.

Atkins, P.W. (1986). *Physical Chemistry*, 3rd edn. Oxford University Press.

Drever, J. (1988). *The Geochemistry of Natural Waters*, 2nd edn. Englewood Cliffs, NJ: Prentice Hall.

Fergusson, J.E. (1982). *Inorganic Chemistry and the Earth*. Oxford: Pergamon Press.

Laidler, K.J. (1987). *Chemical Kinetics*, 3rd edn. New York: Harper and Row.

Lewis, G.N & Randall, M. (1961). *Thermodynamics*, 2nd edn. (revised by K.S. Pitzer & L. Brewer). New York: McGraw-Hill.

Malone, L.J. (1989). *Basic Concepts of Chemistry*, 3rd edn. New York: John Wiley & Sons.

Manahan, S. (1984). *Environmental Chemistry*, 4th edn. Boston: Willard Grant Press.

Riley, J.P. & Skirrow, G. (eds.) (1975). *Chemical Oceanography*, 2nd edn., vol. 1. London: Academic Press.

Segal, B.G. (1989). *Chemistry, Experiment and Theory*, 3rd edn. New York: John Wiley & Sons.

3

○ ○ ○ ○ ○ ○ ○ ○ ○ ○ ○ ○ ○ ○ ○ ○ ○ ○ ○

Chemistry of the elements

3.1   Natural abundance of the elements

Our knowledge about the elemental composition of the earth is restricted to the composition of the crust, atmosphere and oceans. Although the mantle and core represent over 99% of the earth's mass, their elemental compositions are not accurately known. Hence natural abundances of elements quoted in Table 3.1 refer only to the crust. Since the crust (35 km average thickness under continents and 10 km under oceans) is not homogeneous and different parts of the earth contain different minerals, these values are averages of a large number of estimations. The relative homogeneity of ocean waters (Tables 5.3 and 5.4) and the atmosphere (Table 3.2) renders their compositions far less variable for major components.

Oxygen is the major element in the crust, with silicon coming second in abundance. Formation processes may have enriched the crust with certain elements in comparison with the mantle and core. The water and air masses are very different in composition from the crust, and, although elemental abundances may vary slightly with water depth or atmospheric height, the overall relative proportions of the major elements remain almost constant in the oceans and in the lower and middle atmosphere.

Although many problems in environmental chemistry are related to the abundance of elements in the environment, the chemical behaviour and properties of elements are largely independent of their abundances. These properties are, however, dependent upon the atomic mass and electronic configuration of an element, whilst similarities in chemical behaviour between groups of elements have been demonstrated after arranging the elements in order of their atomic numbers. This arrangement in rows is called the periodic table (Table 1.1) and has been introduced in Section 1.1.6.

Table 3.1. *Abundance of elements in the earth's crust (grams per metric ton)*

O	466 700	Li	65	Ta	2.0
Si	277 200	N	46	Br	2.0
Al	81 300	Ce	46	Sc	1.5
Fe	50 000	Sn	40	Eu	1.1
Ca	36 000	Co	40	In	1.0
Na	28 300	Y	28	Sb	1.0
K	25 900	Nd	24	Tb	0.9
Mg	20 900	Nb	24	Lu	0.8
Ti	4400	La	18	Hg	0.5
H	1400	Pb	16	Tl	0.3
P	1200	Ga	15	I	0.3
Mn	1000	Mo	15	Bi	0.2
F	800	Th	12	Tm	0.2
S	520	B	10	Cd	0.2
Cl	480	Cs	7	Ag	0.1
Ba	430	Ge	7	Se	0.09
C	320	Sm	6.5	At	0.04
Rb	310	Gd	6.4	Pd	0.01
Zr	220	Be	6.0	Pt	0.005
Cr	200	Pr	5.5	Au	0.005
Sr	150	As	5.0	He	0.003
V	150	Hf	5.0	Te	0.002
Ni	100	Dy	4.5	Rh	0.001
Zn	80	U	4.0	Re	0.001
Cu	70	Yb	2.7	Ir	0.001
W	69	Er	2.5	Os	0.001
				Ru	0.001

Table 3.2. *Gaseous composition of unpolluted dry air*

Component	Chemical form	Concentration (ppm by volume)
Nitrogen	N_2	780 840
Oxygen	O_2	209 460
Argon	Ar	9340
Carbon dioxide	CO_2	340
Neon	Ne	18
Helium	He	5.2
Methane	CH_4	1.6
Krypton	Kr	1.1
Hydrogen	H_2	3.0
Nitrous Oxide	N_2O	0.3
Xenon	Xe	0.09

We shall now examine relevant chemical properties of the elements in relation to their position in the periodic table.

3.2   Main block elements. Part 1: hydrogen, groups IA, IIA, IIB, IIIB, IVB

3.2.1 General properties and periodic trends

The periodic table (Table 1.1) is highly successful in demonstrating the fact that elements with the same outer electron configuration have similar properties. It must be pointed out, however, that no two elements are the same or show exactly the same chemical behaviour. Although they are arranged in groups with the same outer electron configuration, but different atomic number, their bulk chemical and physical properties (e.g. melting point, boiling point, conductivity, etc.) and atomic properties (e.g. atomic radius, electronegativity, etc.) vary on ascending or descending a particular group. Obviously, these properties also vary on traversing long periods (i.e. going across the periodic table).

The main group elements (Figure 3.1) are also called the s- and p-block elements because only their outer s and p electrons are involved in bonding. They are subdivided into nine groups as follows. Group IA or alkali metals, IIA or alkaline earths, IIB (zinc, cadmium and mercury), IIIB or the boron group, IVB the carbon group, VB the nitrogen group, VIB the oxygen group, VIIB the halides, and group 0, the noble gases. Hydrogen is a group of its own because it does not have properties compatible with any single group. Elements can also be subdivided into three categories: metals, non-metals and metalloids (previously known as semiconductors, a classification originating from the extent of their ability to conduct electricity). Other physical and chemical properties comply with this classification further distinguishing metals from non-metals or metalloids, although the initial criterion for classification was their conductivity (Figure 3.1).

Initial classification of the main group elements is further substantiated by studying periodic trends of some of their important properties. The melting point is defined as the temperature at which pure solid is in equilibrium with pure liquid at one atmosphere pressure. It mainly depends on the type and strength of bonding in the bulk element. Alkali metals, held by weak metallic bonding, have mid-range melting points. The melting point decreases as the group is descended as the size of the atoms increases and the bonding gets weaker. By comparison, the carbon group elements with covalent bonds holding together atoms in the bulk have high melting points, again decreasing in value as the group is descended. Halogens, held by weak intermolecular forces known as van der Waal's forces, have very low melting points and exist as gases. The boiling point is defined as the temperature at which a pure liquid or solid is in equilibrium with its vapour at one atmosphere pressure. Periodic trends are similar to those for melting point.

Bulk metals, in which the bonding electrons are widely distributed around the nuclei, conduct electricity much more efficiently than non-metals, in which electrons are localised in covalent bonds. Similar patterns are observed for thermal conductivity, with non-metals being poor conductors. Metallic character increases as groups are descended and as periods are traversed from right to left. Both electrical and thermal conductivity follow these trends.

Oxidation states (see Section 1.2.2) in traversing a period show trends associated with the gain or loss of electrons in achieving noble gas configuration of $1s^2$ or ns^2np^6 ($n = 1, 2, 3 \ldots$). Hence elements either lose or gain electrons so that this configuration can be achieved. In doing so they form bonds with other elements. It is therefore important to know or predict these oxidation states, although many elements can

Figure 3.1. Metallic and non-metallic classification of the main group elements.

metals			metalloids		non-metals			
IA	IIA	IIB	IIIB	IVB	VB	VIB	VIIB	0
Li	Be		B	C	N	O	F	Ne
Na	Mg		Al	Si	P	S	Cl	Ar
K	Ca	Zn	Ga	Ge	As	Se	Br	Kr
Rb	Sr	Cd	In	Sn	Sb	Te	I	Xe
Cs	Ba	Hg	Tl	Pb	Bi	Po	At	Rn

exist in more than one of them, depending on their particular chemical environment (Figure 3.2).

Atomic radii of elements are defined as distances from the nucleus at which they exclude other atoms. Because the nucleus of each atom is very small in comparison with its overall volume, electrons dictate its size. Atomic radii then depend on the kind of bonding these electrons are engaged in, and it is important to define the atomic radii accordingly (ionic radii, covalent radii, metallic radii, and van der Waal's radii). In traversing a row, from left to right, atomic radii decrease because of the increase in effective nuclear charge, i.e. the charge in the nucleus increases but addition of one electron, in going across the period, does not shield the other electrons from increased attraction from the nucleus and the electron cloud is drawn in more tightly. The rate of such decrease, however, is diminishing as the atoms become heavier. For example, the decrease from lithium to fluorine is much more pronounced than the decrease from rubidium to iodine. On descending a group atomic radii increase because, although nuclear charges increase, they cannot counteract the increase in volume from addition of whole shells of electrons. Again this rate of increase in size is diminishing as atomic numbers increase. Hence the increase in size is more pronounced on descending group IA than group VIIB. All periodic trends described above are true for all kinds of atomic radii provided similar radii are compared (e.g. ionic radii with ionic radii).

Figure 3.2. Common oxidation states of main group elements.

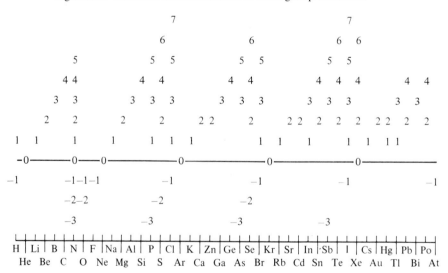

The ionisation energy of an element is a measure (in kJ mol^{-1} at 0 K) of how firmly the nucleus of a gaseous atom or ion attracts or holds its outer electrons. Periodic trends of first ionisation energies (energies for removal of the first electron from an atom) are linked to atomic radii by virtue of the effective nuclear charge (i.e. the attraction of the nucleus experienced by the outer electrons). The stronger the effective nuclear charge the more difficult it is to extract an electron from the element. Anomalies occur because it is more difficult to extract electrons from filled orbitals (e.g. helium in comparison to hydrogen and lithium) and from half-filled shells (e.g. nitrogen in comparison to carbon and oxygen, Figure 3.3).

Electronegativity is a measure of the tendency of elements to become negatively charged. The most common scale used in quantifying such a

Figure 3.3. First ionisation energies of main group elements.

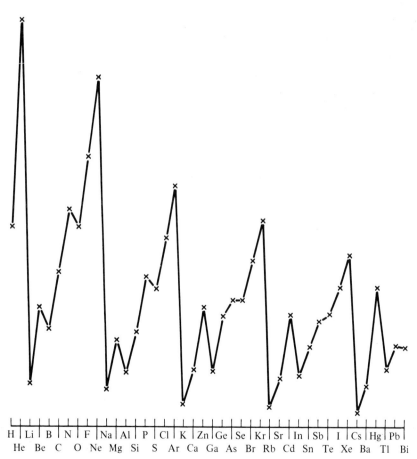

tendency is the Allred–Rochow (A/R) scale, which is based on the measurement of effective nuclear charge. On traversing a row from left to right electronegativity increases as effective nuclear charge increases. Again on descending a group it decreases. Thus, one of the least electronegative elements is caesium and the most electronegative is fluorine. This is reflected in the chemical properties of the elements: caesium readily loses an electron to form a Cs^+ ion, whilst fluorine readily gains one to give F^- (see Table 1.2).

3.2.2 Hydrogen

Hydrogen cannot be classified in any other group of elements in the periodic table because of its unique properties, and therefore forms a group of its own. It is the simplest of elements, consisting of a singly charged nucleus with one electron in the 1s orbital. However, it has three isotopes: ^1H, or hydrogen; ^2H, or deuterium (D, nucleus with one proton and one neutron); and ^3H or tritium (T, nucleus with one proton and two neutrons). The latter two isotopes are found at only very low abundance in the natural environment and tritium is radioactive. Some estimates suggest that hydrogen is the most abundant element in the universe, which is estimated to contain 92% hydrogen. It is certainly extremely important for life as known to us because it forms oxygen hydride (or water). Although it forms hydrides with metals, they are not very stable in air, decomposing to give molecular hydrogen, e.g.

Figure 3.4. The structure of hydrogen bonds in water.

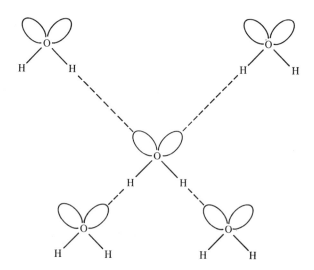

$$SnH_4 \rightarrow Sn + 2H_2$$

The most important compounds are the ones in which hydrogen bonds covalently (e.g. in organic compounds of carbon, in ammonia and water), and ionic compounds containing H^+ ions; the strong acids such as H_2SO_4.

The human body contains *ca.* 65% water, and life as we know it would not exist without it. Physical properties of water do not follow expected trends of other hydrides because of hydrogen bonding. Hydrogen, although bonded to one oxygen atom in a water molecule, is attracted to the electron pair of another oxygen atom (see Figure 3.4). This attraction is electrostatic, and although not formally a bond is sufficient to elevate melting and boiling points to values higher than theory would predict (e.g. predicted boiling point *ca.* 75°C; actual 100°C). Because of its polar nature, water is a very good solvent for ionic compounds, and natural waters can contain very high concentrations of dissolved solids. Sea water contains *ca.* 35 g l^{-1} dissolved solids, mostly Na^+ and Cl^- ions, but also other ions, some of which can be extracted commercially (e.g. Br^-). Fresh waters and water consumed domestically contain a variety of other dissolved material, some of which may be added (e.g. F^-), or naturally present (e.g. Ca^{2+} and Mg^{2+} ions, which give rise to water hardness).

3.2.3 Group IA: the alkali metals

The most abundant alkali metals are sodium and potassium, found mainly as their chloride salts. The term alkali metals originated from their ability to increase the OH^- concentration of water, hence increasing its pH, e.g.

$$2Na + 2H_2O \rightarrow 2NaOH + H_2\uparrow$$

Sodium and its compounds are very important in nature. Sodium chloride (NaCl or common salt) contributes to the salinity of sea water and may play an important part in the cycles of acids in the atmosphere, e.g.

$$HNO_3(g) + NaCl \rightarrow NaNO_3 + HCl(g)$$

Both sodium and potassium are used industrially in a variety of processes, and K_2SO_4 is used as a fertiliser. Sodium tripolyphosphate, produced industrially, is used in detergents to complex Ca^{2+} and Mg^{2+} ions which otherwise contribute to the hardness of water. At modest doses sodium and potassium are not toxic to mammals but have physiological functions; both are involved in transmission of nerve pulses through living cells. Lithium is of no special environmental significance.

3.2.4 Group IIA: the alkaline earth metals

Beryllium is not very abundant in nature but is of some commercial importance. It is very toxic, leading to degeneration of the lung when inhaled. Magnesium, calcium, strontium and barium are widely distributed elements with large deposits of limestone ($CaCO_3$) and dolomite ($CaCO_3 \cdot MgCO_3$) in the earth's crust. Carbonates and bicarbonates of calcium and magnesium contribute to the hardness of water due to leaching from rock by water acidified by the presence of carbon dioxide (see Section 5.2.2):

$$CaCO_3 + CO_2 + H_2O \rightarrow Ca(HCO_3)_2 \qquad \text{(calcium bicarbonate)}$$

Magnesium carbonate behaves in a similar manner. Magnesium is essential to all organisms and present in all chlorophylls. Like all the other group IIA elements it forms ionic compounds in the solid state with an oxidation number of $+2$. Barium and radium are less abundant, but of generally similar chemistry. All isotopes of radium are naturally radioactive.

3.2.5 Group IIB: zinc, cadmium and mercury

The group IIB metals have an ns^2 configuration of outer electrons, and, although they contain d and f electrons, these are not involved in bonding. Hence their chemical behaviour and properties position them with the s and p block elements. They all form compounds in their $+2$ oxidation state whilst mercury also exhibits the $+1$ oxidation state. They are used in metallurgical processes, and smelters are among the main pollution sources contributing to environmental zinc and cadmium burdens. They are all relatively volatile metals, readily lost from pyro-metallurgical processes. Mercury is of particularly high vapour pressure and enters the atmosphere by degassing of the earth's crust. Zinc may be used in fertilisers for some zinc-deficient soils because some known pathological conditions are associated with dietary deficiencies in zinc. Cadmium pollution originates from smelters, incineration of plastics and cadmium pigments, fossil fuels, electroplating, metallurgical processes, etc. Recently, use on land of sewage sludge with a high concentration of metals, including cadmium, as fertilisers has caused concern. Cadmium is generally highly available to plants although availability depends on the composition and presence of complexing agents in the soil.

Mercury is still widely used commercially because of its unique physical properties. As a liquid metal at room temperature it has been used in thermometers, mobile cathodes, diffusion pumps, as well as in amalgams (alloys of mercury with other metals). Both inorganic and organic mercury species are highly toxic to a large number of living organisms and adversely affect the central nervous system in mammals. Because of the toxicity, the organic compound RHgX (R = phenyl or alkyl, see Sections 3.3 and 3.4; X = counter-ion) have been utilised as pesticides and fungicides. As the metal to carbon bond is relatively stable in water and air, these compounds can accumulate in the food chain. Concentrations can increase significantly at higher trophic levels, including man and other mammals at the top of the food chain. This process is termed 'bioamplification'. There are known cases of accidental organic mercury poisonings where fish or grain containing organomercurials have been ingested by humans. Although Hg^{2+} can be biomethylated in the environment (addition of one or two CH_3^- groups from biochemicals such as vitamin B_{12}) the yields of such reactions are low and poisoning from biomethylated products has not been recorded:

$$HgX_2 + CH_3CoB_{12} \xrightarrow{H_2O} CH_3HgX + H_2OCoB_{12}^+ + X^-$$
$$\text{(methylcobalamin)}$$

The use of organomercury fungicides and slimicides (for prevention of slime forming in effluents of wood pulp industries) is diminishing because of recognition of the organomercury compounds' persistence in the environment.

Mercury(I) compounds are known but most are unstable and susceptible to disproportionation reactions (two molecules of a metallic compound produce two molecules of the metal at oxidation states different to the original one):

$$Hg(I)_2Cl_2 \rightarrow Hg(0) + Hg(II)Cl_2$$

3.2.6 Group IIIB: the boron group

In the previous groups of the main block elements only s electrons were used in bonding. From group IIIB to VIIB, p electrons are also involved in bonding, giving rise to more than one stable oxidation state (Figure 3.2). The metallic character of group IIIB increases as it is descended, with boron showing non-metallic chemical properties (but semiconducting electricity when pure) and elements from aluminium to thallium being metals.

Aluminium is the third most abundant element in the environment, occurring mainly in rocks and soils in the form of clays and other minerals. It has been postulated that water acidity (from acid rain) causes dissolution of aluminium which may be toxic to fish. Section 5.4 describes the chemical processes involved. There have also been suggestions (as yet unproven) that aluminium can be a cause of Alzheimer's Disease (a form of pre-senile dementia) in humans.

3.2.7 Group IVB: the carbon group

All elements of the group exhibit an oxidation number of $+4$ and/or $+2$. Carbon, although not the most abundant element in the earth's crust, is the basis of life as we know it. Its compounds containing hydrogen, oxygen, nitrogen and to a lesser extent other elements, are usually studied separately in the branch of chemistry called 'organic chemistry'. However, some important environmentally occurring compounds are not classified as organic compounds. Carbon dioxide (CO_2) is a product of plant metabolism, combustion of organic compounds (fossil fuels), fermentation processes, etc. As discussed in Chapter 2, CO_2 dissolution in natural water systems establishes the following equilibria:

$$CO_2(g) \rightleftharpoons CO_2(aq) + H_2O \rightleftharpoons H_2CO_3 \rightleftharpoons H^+ + HCO_3^- \rightleftharpoons 2H^+ + CO_3^{2-}$$

Combustion of forests and fossil fuels has ensured that atmospheric levels of CO_2 have increased this century. There is concern that because CO_2 plays an important role in the radiation balance of the earth, increased concentrations will lead to heating of the lower atmosphere. This anticipated climatological change due to the 'greenhouse' effect is of uncertain magnitude at present, partly due to the possibility that other effects, e.g. increased cloud covering, will cause cooling.

Incomplete combustion of fossil fuels gives rise to carbon monoxide, CO, which is highly toxic to mammals because it binds haemoglobin in the blood, inhibiting oxygen exchange.

Silicon is the second most abundant element in the earth's crust, after oxygen (27.7%). It occurs widely in rock such as granite, in silicates and as silica, SiO_2, which occurs as the mineral quartz in sand and sandstone. It is not appreciably toxic but when inhaled in the form of SiO_2 can injure the mammalian lung due to its irritant properties. Its organic compounds are used in production of polymers known as silicones. Both silicon and germanium are used in the semiconductor industry in a high purity form.

Tin is not considered as a toxic metal, although the organic compounds can be highly toxic, depending on the organic group attached. These organic compounds are used in pvc (polyvinylchloride polymer) as stabilisers and plasticisers. They act by scavenging Cl^- formed by the degradation of pvc and replacing it with bulky alkyl groups, hence stopping propagation of the Cl^- removal. The most commonly known tin compound used as an antifouling agent in the shipping industry is TBTO (bistributyltin oxide; $(C_4H_9)_3Sn$—O—$Sn(C_4H_9)_3$). It is mixed in paints used on the underside of ships. It is slowly released by the action of saline water forming $(C_4H_9)_3SnCl$ which is toxic to marine organisms that foul ships:

$$(C_4H_9)_3Sn\text{—O—}Sn(C_4H_9)_3 + 2NaCl + H_2O \rightarrow 2(C_4H_9)_3SnCl + 2NaOH$$

However, tin compounds are also toxic to shellfish, and their use in shallow waters is currently being limited in some countries. Tin methyl compounds are highly toxic to mammals. Although Sn^{4+} and Sn^{2+} can be methylated by biological processes (unknown biochemical mechanisms) or CH_3^+ ion donors, respectively, the process is inefficient and yields are too low to cause concern:

$$CH_3^+ + Sn^{2+} \rightarrow CH_3Sn^{3+}$$

$$CH_3^- + Sn^{4+} \rightarrow CH_3Sn^{3+} \xrightarrow{CH_3^-} (CH_3)_2Sn^{2+}$$

etc.

Lead, a well known metal from ancient times, is toxic to mammals, affecting the central nervous system. Usage in plumbing is now discouraged due to dissolution in potable waters (plumbosolvency). The use of red lead oxide (Pb_3O_4) and other lead compounds in pigments is also limited, in places where young children may come into contact with it.

The organic compounds of lead (mainly tetraethyllead $(C_2H_5)_4Pb$ and tetramethyllead $(CH_3)_4Pb$) are used as antiknock additives in automobile fuel. They prevent secondary explosions in the combustion chamber, known as 'knock'. They are produced commercially from lead/sodium alloys as follows:

$$4NaPb + 4RCl \rightarrow 4NaCl + R_4Pb + 3Pb \qquad (R = C_2H_5 \text{ or } CH_3)$$

They are toxic compounds to mammals but are almost completely burned to inorganic lead when used in automobile fuel. This process contributes to increased lead concentrations in the environment. In recent years use of alkyllead compounds has declined because of uncertainties about the long-term effects of lead on human health, and because lead poisons vehicle exhaust catalysts installed to abate gaseous air pollutant emissions.

3.3   Organic chemicals

No simple and entirely consistent definition exists for what constitutes an organic, as opposed to inorganic, compound. All organic compounds contain carbon, and the vast majority also contain hydrogen, bonded to carbon. Most chemists exclude certain carbon containing compounds, such as metal carbonates, carbides and cyanides from their definition of organic compounds.

Carbon exhibits a very diverse chemistry and forms a wide range of compounds. This behaviour arises in part because carbon is tetravalent and can thus form four single bonds to other atoms. However, bonding may occur through sp, sp^2 on sp^3 hybridisation (as outlined in Section 1.2.4). Carbon can form multiple bonds to several other elements. Also, carbon can form chains of enormous variability in length, bonding characteristics and elemental composition, in a manner of which no other element is capable, although silicon exhibits some similar chemistry. Finally, carbon atoms are involved in a wide range of natural and man-made polymers. A *polymer* is a large molecule consisting of many units of one or more monomer molecules bonded together. Starch and cellulose are natural organic polymers. Carbon atoms can form a wide range of compounds, even when in combination with only one other element such as hydrogen.

Most organic compounds belong to groups known as *homologous series*. These series each have in common a *functional group*, that is a grouping of atoms which is readily identifiable and confers characteristic chemical properties upon the molecule but have different numbers of carbon atoms attached. Accordingly, the nomenclature of organic compounds is derived from such functional groups together with the number of carbon atoms linked in a chain within the molecule. Functional groups and compounds combining one carbon atom (i.e. C_1) will be identified using the prefix methyl or meth-; C_2 by ethyl or eth-; C_3 by propyl or prop-; etc. The naming of compounds will be considered further in the following sections. A general distinction made in classifying organic compounds is whether or not they contain a benzene ring. *Aliphatic* compounds consist of open chains of carbon atoms, while *aromatic* compounds are distinguished by the presence of one or more closed benzene rings. A compound containing a ring of carbon atoms other than a benzene ring is termed *alicyclic*, whilst if one or more atoms in the ring is of an element other than carbon, the compound is termed *heterocyclic*.

3.3.1 Alkanes

These are the simplest and least reactive hydrocarbons (compounds of sp³-hybridised carbon and hydrogen). They are said to be *saturated*, meaning that they contain no double or triple bonds capable of accepting further hydrogen atoms (cf. the *unsaturated* alkenes and alkynes).

The common structural formula for alkanes is:

$$C_nH_{2n+2}$$

where n takes any integer value from 1 to >20. The simpler members of the series appear in Table 3.3 Each successive compound is larger than the one before by the addition of one CH_2 group. The prefix '*n*-' in Table 3.3 is indicative of linear structure. From C_4H_{10} onwards, branched chain structures are possible, e.g. for butane

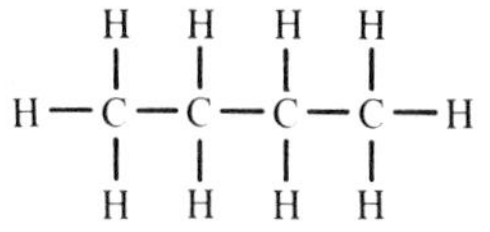

n-butane

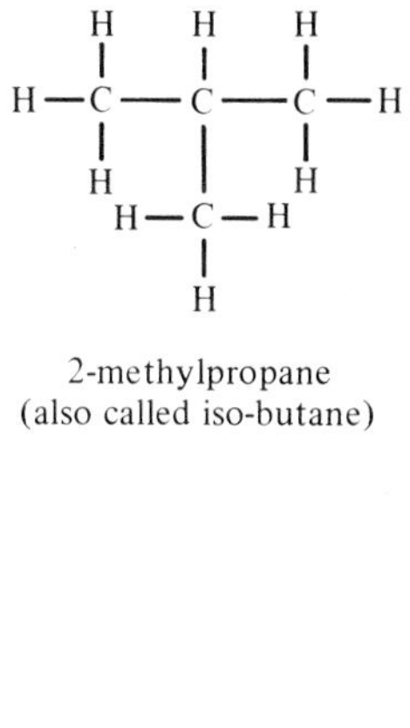

2-methylpropane
(also called iso-butane)

or for pentane

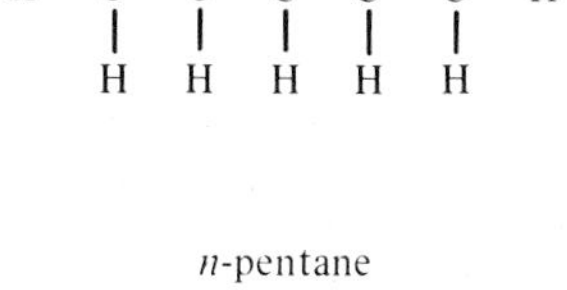

n-pentane

2,2-dimethylpropane
(also called neo-pentane)

2-methylbutane
(also called iso-pentane)

These different forms are called *structural isomers* and represent entirely different compounds, with different physical and chemical properties.

Table 3.3. *Common alkanes*

Name	Structural formula	M.p. (°C)	B.p. (°C)
Methane		−184	−164
Ethane		−172	−88.6
Propane		−190	−44.5
n-Butane		−135	−0.5
	Molecular formula		
n-Pentane	C_5H_{12}	−129	36
n-Hexane	C_6H_{14}	−94	68.9
n-Heptane	C_7H_{16}	−90.6	98.4
n-Octane	C_8H_{18}	−56.8	125.7
n-Nonane	C_9H_{20}	−51	150
n-Decane	$C_{10}H_{22}$	−30	174

If a hydrogen atom is removed from an alkane, the resulting free radical is termed an alkyl group. Thus

CH_3 methyl, C_2H_5 ethyl, C_3H_7 propyl, etc.

Occurrence of alkanes in the environment

The lowest member of the series, methane, is a ubiquitous component of the atmosphere, having a global background concentration of *ca.* 1.6 ppm (v/v). It is formed naturally from biological decay processes: the decomposition, for example, of sewage sludges gives rise to abundant quantities of a gas rich in methane. The major component of natural gas is also methane. Other low molecular weight alkanes are present in these gases, but at far lower concentrations than methane.

The alkanes which are liquid at normal temperatures (C_5H_{12} and above, see Table 3.3) are abundant in petroleum and are described in Section 3.4.1. Their use in petrol and diesel fuels leads to their presence in polluted air due to evaporation and incomplete combustion.

Chemical properties of alkanes

Because of their saturation and lack of functional groups, alkanes are chemically very unreactive. Methane is especially so and has an atmospheric lifetime of the order of years. The only reactions of alkanes of environmental significance are with reactive free radicals such as hydroxyl (see Section 5.1.1) which may abstract a hydrogen atom to form an alkyl free radical, itself a very reactive species. Thus, for example,

$$CH_4 + OH \rightarrow CH_3 + H_2O \qquad (3.1)$$

3.3.2 Alkenes

The characteristic functional group within the alkenes is the carbon to carbon double bond in which carbon is sp^2 hybridised. Thus ethene, the simplest member of the series is:

$$\begin{array}{c} H \\ \diagdown \\ H \diagup \end{array} C = C \begin{array}{c} \diagup H \\ \diagdown \\ H \end{array}$$

Higher molecular weight alkenes are formed by the inclusion of alkyl groups (Table 3.4). The common structural formula of the alkenes is C_nH_{2n}, and structural isomerism occurs for compounds where $n \geqslant 4$. This is illustrated in Table 3.4, which shows that two types of isomerism are possible. The first is due to a different arrangement of carbon atoms (i.e. branched chain or straight chain in this case), and the second due to the position of the double bond in the carbon chain. The position of the double bond is counted from the nearest end of the molecule, hence leading to the names but-1-ene and but-2-ene:

$$CH_2\!=\!CHCH_2CH_3 \qquad CH_3CH\!=\!CHCH_3$$
$$\text{but-1-ene} \qquad\qquad \text{but-2-ene}$$

One further complication arises from the fixed planar nature of the molecule. In the alkane ethane the carbon atoms may rotate freely about the single bond. However, in ethene no rotation about the double bond is possible and the molecule is essentially fixed in one plane. This gives rise to *geometric isomerism*, e.g. for but-2-ene

cis-but-2-ene   *trans*-but-2-ene

Thus the prefixes *cis-* and *trans-* are used to denote, respectively, compounds with groups on the same, and on opposite sides of the molecule. This type of isomerism does not occur in alkanes where rotation about the central C—C bond allows enormously rapid inter-change of structures.

Table 3.4. *Common alkenes*

Systematic name	Common name	Structure	B.p. (°C)
Ethene	Ethylene	$CH_2\!=\!CH_2$	−105
Propene	Propylene	$CH_3\!-\!CH\!=\!CH_2$	−48
But-1-ene		$CH_3\!-\!CH_2\!-\!CH\!=\!CH_2$	−6.1
But-2-ene		$CH_3\!-\!CH\!=\!CH\!-\!CH_3$	+1
2-Methylpropene		$CH_3\!-\!\underset{\underset{CH_3}{\textstyle\mid}}{C}\!=\!CH_2$	−6.6

Occurrence of alkenes in the environment

The simplest member of the group, ethene, is a plant hormone, produced naturally within living plants. Other compounds of higher molecular weight containing alkene groupings occur widely as natural products in nature, and include natural rubber, isoprene, and some terpenes from trees and carotenoid pigments.

Although not present in motor fuels to any extent, alkenes are important as products of incomplete combustion of gasoline, from which they enter the atmosphere in considerable amounts.

Chemical properties of alkenes

Because of the presence of the unsaturated double bond, alkenes are far more reactive than alkanes. Many chemical reagents can add to the alkene double bond, and this process can be of importance in photochemical smog (Section 5.1.2) since it generally leads to cleavage of the molecule with formation of one or more reactive radical species. Ozone can insert itself in the double bond. Thus, for example, with but-2-ene and ozone

ozonide intermediate

$$(3.2)$$

This reaction is termed oxidative cleavage.

3.3.3 Alkynes

This is the third homologous series of hydrocarbons and contains two sp hybridised carbons with a carbon–carbon triple bond. Thus the structural formula is C_nH_{2n-2}, and the simpler members of the series appear in Table 3.5.

Alkynes occur only rarely in nature, alkynic bonds appearing in some complex fungal metabolites. Ethyne (known commonly as acetylene) is formed in incomplete gasoline combustion and is emitted from cars. It is

Table 3.5. *Lower molecular weight alkynes*

Name	Structure	B.p. (°C)
Ethyne (acetylene)	$H-C\equiv C-H$	-84
Propyne	$CH_3-C\equiv C-H$	-27.5
But-1-yne	$C_2H_5-C\equiv C-H$	$+8.3$
But-2-yne	$CH_3-C\equiv C-CH_3$	$+27.5$

surprisingly unreactive in air and is far more persistent in a polluted air mass than the alkenes or higher molecular weight alkanes.

3.3.4 Alcohols

This group of compounds is characterised by the presence of the hydroxyl functional group —OH covalently bonded to carbon. The simpler members of the series are listed in Table 3.6.

Alcohols are widely used in industry and commonly appear as 'organic micropollutants' of water (i.e. organic pollutant present at low concentrations, typically $\mu g\,l^{-1}$ or less). The alcohol glycerol is important naturally (see later).

3.3.5 Aldehydes and ketones

Both aldehydes and ketones contain the carbonyl functional group C=O. In aldehydes this is bonded to one hydrogen atom and one alkyl or other group (denoted R), whilst in a ketone it is bonded to two alkyl groups:

$$\begin{array}{c} R \\ \diagdown \\ \diagup \\ H \end{array} C=O \qquad \begin{array}{c} R \\ \diagdown \\ \diagup \\ R' \end{array} C=O$$

     aldehyde   ketone

The simpler members of both series are listed in Tables 3.7 and 3.8.

Many aldehydes and ketones have important industrial uses and hence appear as organic micropollutants in water. These compounds are also formed in the atmosphere by oxidation of hydrocarbons by ozone, atomic oxygen or hydroxyl (e.g. reaction 3.2).

It is appropriate to mention two chemical reaction processes

Table 3.6. *Common alcohols*

Systematic name	Common name	Structure	B.p. (°C)
Methanol	Methyl alcohol	$H-\overset{\displaystyle H}{\underset{\displaystyle H}{C}}-OH$	64.1
Ethanol	Ethyl alcohol	$H-\overset{\displaystyle H}{\underset{\displaystyle H}{C}}-\overset{\displaystyle H}{\underset{\displaystyle H}{C}}-OH$	78.5
Propan-1-ol	n-Propyl alcohol	$H-\overset{\displaystyle H}{\underset{\displaystyle H}{C}}-\overset{\displaystyle H}{\underset{\displaystyle H}{C}}-\overset{\displaystyle H}{\underset{\displaystyle H}{C}}-OH$	97.4
Propan-2-ol	iso-Propyl alcohol	$H-\overset{\displaystyle H}{\underset{\displaystyle H}{C}}-\overset{\displaystyle H}{\underset{\displaystyle O-H}{C}}-\overset{\displaystyle H}{\underset{\displaystyle H}{C}}-H$	82.4

Table 3.7. *Low molecular weight aldehydes*

Systematic name	Common name	Structure	B.p. (°C)
Methanal	Formaldehyde	$H-CHO$	−21
Ethanal	Acetaldehyde	CH_3-CHO	21
Propanal	Propionaldehyde	C_2H_5-CHO	48
Butanal	n-Butyraldehyde	$CH_3CH_2CH_2-CHO$	75
2-Methyl propanal	Isobutyraldehyde	$(CH_3)_2CH-CHO$	64

Table 3.8. *Low molecular weight ketones*

Systematic name	Common name	Structure	B.p. (°C)
Propanone	Acetone	CH_3COCH_3	56.2
Butanone	Ethyl methyl ketone	$CH_3CH_2COCH_3$	79.6
Pentan-2-one	Methyl n-propyl ketone	$CH_3COCH_2CH_2CH_3$	102
Pentan-3-one	Diethyl ketone	$CH_3CH_2COCH_2CH_3$	101.5
3-Methylbutan-2-one	Isopropyl methyl ketone	$\underset{CH_3}{\overset{CH_3}{>}}CHCOCH_3$	94

involving these compounds. In the human body ethanol is oxidised, first to ethanal, and then to ethanoic acid (acetic acid):

$$CH_3CH_2OH \rightarrow CH_3CHO \rightarrow CH_3COOH$$

These reactions also form the basis of the production of vinegar, and are quite rapid in the presence of appropriate enzymes.

3.3.6 Carboxylic acids and esters

As noted above, oxidation of ethanal (an aldehyde) yields ethanoic acid (a carboxylic acid). Carboxylic acids contain the carboxyl functional group

$$-C\underset{OH}{\overset{O}{\big/\!\!\big/}}$$

and examples appear in Table 3.9.

The carboxylic acids are partially dissociated in aqueous solution

$$CH_3COOH + H_2O \rightleftharpoons CH_3COO^- + H_3O^+ \qquad (3.3)$$

They are only weak acids giving a very low degree of dissociation

$$K_a = \frac{[H_3O^+][CH_3COO^-]}{[CH_3COOH]} = 1.85 \times 10^{-5} \text{ mol dm}^{-3} \text{ at } 25°C \qquad (3.4)$$

Carboxylic acids have important industrial uses and occur widely as organic micropollutants of waters.

Esters are formed by reaction of a carboxylic acid and an alcohol:

$$R-C\underset{OH}{\overset{O}{\big/\!\!\big/}} \quad + \quad R'OH \quad \rightleftharpoons \quad R-C\underset{OR'}{\overset{O}{\big/\!\!\big/}} \quad + H_2O \qquad (3.5)$$

carboxylic acid   alcohol   ester

Table 3.9. *Low molecular weight carboxylic acids*

Systematic name	Common name	Formula	M.p. (°C)	B.p. (°C)
Methanoic acid	Formic acid	HCOOH	8.4	100.5
Ethanoic acid	Acetic acid	CH_3COOH	16.7	118.2
Propanoic acid	Propionic acid	CH_3CH_2COOH	−19.7	141.4
Butanoic acid	n-Butyric acid	$CH_3CH_2CH_2COOH$	−8	162
2-Methyl propanoic acid	Iso-butyric acid	$\begin{array}{l}CH_3\\ \!\!\diagdown\!\!CHCOOH\\ CH_3\!\!\diagup\end{array}$	−47	154

The ester product of reaction of ethanoic acid (acetic acid) and ethanol is called ethyl ethanoate, or, more commonly, ethyl acetate. Esters are noted for their fruity smells and find use as industrial reagents and solvents. *Lipids* are naturally occurring esters of long-chain carboxylic acids, known as fatty acids. These are described in more detail in Section 3.4.2.

3.3.7 Amines

Amines are organic derivatives of ammonia and may be primary, secondary or tertiary dependent upon the number of alkyl groups (R) contained:

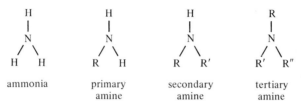

It is possible to add a fourth alkyl group to ammonia, the nitrogen then bearing a positive charge. The resultant compound is called a quaternary ammonium salt. The lower members of the series are listed in Table 3.10. The $-NH_2$ functional group in primary amines is called an *amino group*.

Table 3.10. *Lower molecular weight amines*

Systematic name	Common name	Formula	B.p. (°C)
Primary amines	*Primary amines*		
Aminomethane	Methylamine	CH_3NH_2	−6
Aminoethane	Ethylamine	$C_2H_5NH_2$	16.6
1-Aminopropane	n-Propylamine	$CH_3CH_2CH_2NH_2$	49
2-Aminopropane	Isopropylamine	$(CH_3)_2CHNH_2$	32
1-Aminobutane	n-Butylamine	$CH_3CH_2CH_2CH_2NH_2$	78
Secondary amines			
Dimethylamine		$(CH_3)_2NH$	7
Methylethylamine		$C_2H_5NHCH_3$	35
Diethylamine		$(C_2H_5)_2NH$	55.5
Tertiary amines			
Trimethylamine		$(CH_3)_3N$	4
Triethylamine		$(C_2H_5)_3N$	89.4
Quaternary ammonium salt			
Tetramethyl-ammonium chloride		$[(CH_3)_4N]^+Cl^-$	Solid; decomposes above 230°

Amines occur as micropollutants in water and can also be produced naturally. They are perceptible by smell at extremely low levels of concentration and have characteristic 'fishy' odours.

3.3.8 Aromatic compounds

Aromatic compounds are those which contain the very stable benzene ring. This may be represented as a mixture (termed a resonance hybrid) of two structures (termed canonical forms)

In fact the double bond electrons (termed π electrons) are delocalised, i.e. spread around the whole ring, and ignoring the hydrogens the structure is often represented as in the above right. This delocalisation of electrons renders benzene far more chemically stable than a linear trialkene.

There are many important derivatives of benzene. Some hydrocarbon derivatives are listed in Table 3.11. The convention for naming is illustrated by:

| toluene | (1,2-dimethyl benzene) ortho-xylene | (1,3-dimethyl benzene) meta-xylene | (1,4-dimethyl benzene) para-xylene |

It is possible to substitute other functional groups into the aromatic ring. The substitution of a hydrogen atom by a hydroxyl group leads to the formation of phenol. Methyl phenols are called cresols. Phenols are weakly acidic due to dissociation of the hydroxyl group hydrogen.

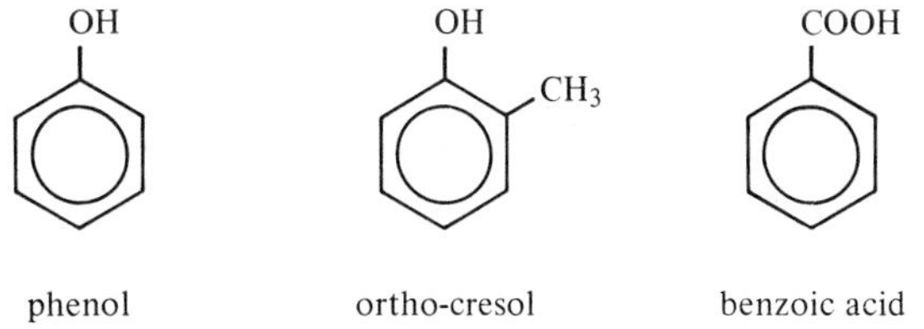

phenol   ortho-cresol   benzoic acid

(3.6)

Introduction of a carboxyl group leads to formation of benzoic acid.

Aromatic hydrocarbons such as benzene, toluene and the xylenes are present in crude oil and in the gasoline (petrol) derived from it. They are emitted into the atmosphere during incomplete petrol combustion. They are used as industrial chemicals, as are many other aromatic compounds, and occur widely as micropollutants of water.

One type of aromatic compound commonly observed in water is the alkyl benzene sulphonate detergent. It is prepared by reaction of an alkyl benzene such as dodecyl benzene with sulphur trioxide to form dodecyl benzene sulphonic acid. Neutralisation with sodium hydroxide gives the alkyl benzene sulphonate (ABS) sodium salt, which is the active detergent in many washing powders.

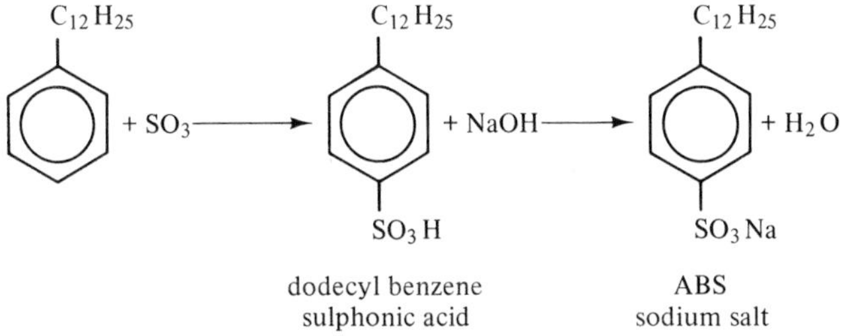

dodecyl benzene
sulphonic acid

ABS
sodium salt

Table 3.11. *Common aromatic hydrocarbons*

Systematic name	Common name	M.p. (°C)	B.p. (°C)
Benzene	Benzene	5.5	80
Methyl benzene	Toluene	−95	111
Ethyl benzene	Ethylbenzene	−94	136
1,2-Dimethyl benzene	o-Xylene	−25	144
1,3-Dimethyl benzene	m-Xylene	−47	139
1,4-Dimethyl benzene	p-Xylene	13	138
1,3,5-Trimethyl benzene	Mesitylene	*	165

* Three forms exist, m.p. −44.8, −49.9, −51.8°C.

It is standard practice currently to use only linear chain alkyl benzenes since the detergents formed from these are biodegradable in a sewage treatment works, whereas the branched alkyl chain derivatives are rather resistant to biological breakdown.

There are also aromatic compounds containing two or more fused benzene rings:

naphthalene   anthracene   phenanthrene

Those compounds, which have three or more rings, are called *polynuclear aromatic hydrocarbons* (PAH). Many PAH are formed in combustion processes and are important pollutants of the atmospheric and aquatic environments. Some PAH are strongly carcinogenic (cancer causing) in experimental animals: one of the best known examples is 3,4-benzopyrene, also known as benzo(a)pyrene:

3,4-benzopyrene
(benzo(a)pyrene)

3.4   Organic compounds in the environment

3.4.1 Petroleum

Crude oil (petroleum) consists primarily of three groups of compounds:

(i) Alkanes, of both straight and branched chain, varying in length from methane up to $C_{70}H_{142}$ approximately.

(ii) Cycloalkanes. These are cyclic saturated compounds such as cyclopentane and cyclohexane

In common with the alkanes, these are chemically rather unreactive.

(iii) Aromatic hydrocarbons, such as benzene, toluene and xylene are abundant in some crude oils.

Crude oil is separated into fractions by distillation on the basis of boiling point. Typical fractions are listed in Table 3.12. These fractions may be further refined by distillation, and may be amended by further chemical or physical treatment to enrich specific compounds required for industrial use.

3.4.2 Lipids

Lipids occur naturally as plant and animal products such as animal fat or vegetable oil. In chemical terms, they are esters formed from long-chain fatty acids and an alcohol, usually glycerol. They may be split into their constituent acids and alcohol by hydrolysis with an alkali such as potassium hydroxide, a process known as *saponification*:

$$
\begin{array}{ll}
CH_2OOCR & CH_2OH \quad RCOO^-K^+ \\
CHOOCR' + 3KOH \rightarrow & CHOH + R'COO^-K^+ \qquad (3.7)\\
CH_2OOCR'' & CH_2OH \quad R''COO^-K^+ \\
\text{lipid} & \text{glycerol} \quad \text{fatty acid salts}
\end{array}
$$

Table 3.12. *Typical distillate fractions from crude oil*

Boiling range (°C)	Name	Uses
20–90	light petroleum	solvent
90–120	ligroin	solvent
100–200	petrol, gasoline } naphtha	fuel
200–300	paraffin, kerosene } naphtha	fuel
above 300	gas oil	fuel
above 300	lubricating oil	lubricant
residue	asphalt, bitumen	road construction

The most common fatty acids occurring in lipids are listed in Table 3.13. Vegetable oils contain a preponderance of unsaturated fatty acids. Olive oil has about 80% octadeca-9-enoic acid (oleic acid) amongst its fatty acid content, whilst animal fats contain both saturated and un-saturated acids. Mutton fat contains esters of palmitic (28%), stearic (28%) and oleic (37%) acids.

Saponification of lipids using sodium or potassium hydroxide leads to formation of the sodium or potassium salts of the fatty acids. These are purified to make soap.

3.4.3 Amino acids and proteins

α-Amino acids are distinguished as having an amino group and a carboxylic acid group bonded to a common carbon atom, thereby giving the general structure

$$\underset{H}{\overset{R}{NH_2 - \underset{|}{\overset{|}{C}} - COOH}}$$

Some of the more important compounds are listed in Table 3.14. These acids can bind together by forming *amide* (—NH—) bonds:

$$NH_2 - \underset{H}{\overset{R}{C}} - COOH + NH_2 - \underset{H}{\overset{R'}{C}} - COOH \longrightarrow NH_2 - \underset{H}{\overset{R}{C}} - \underset{O}{\overset{}{C}} - \underset{H}{\overset{}{N}} - \underset{H}{\overset{R'}{C}} - COOH + H_2O \tag{3.8}$$

The resultant unit containing two amino acids still bears carboxyl and amino functional groups, and hence longer chains known as *peptides* may be formed by addition of further units:

$$-NH - \underset{R}{\overset{}{CH}} - \overset{O}{\overset{\|}{C}} - NH - \underset{R'}{\overset{}{CH}} - \overset{O}{\overset{\|}{C}} - NH - \underset{R''}{\overset{}{CH}} - \overset{O}{\overset{\|}{C}} -$$

Long chains involving perhaps fifty or more amino acid units are known as *proteins*. Natural catalysts known as enzymes (Section 2.1.5) consist of protein; (e.g. DNA polymerase, a catalyst in the polymerisation of DNA).

Table 3.13. *Common natural fatty acids*

Systematic name	Common name	Structure	M.p. (°C)
Dodecanoic acid	Lauric acid	$CH_3(CH_2)_{10}COOH$	44
Tetradecanoic acid	Myristic acid	$CH_3(CH_2)_{12}COOH$	58
Hexadecanoic acid	Palmitic acid	$CH_3(CH_2)_{14}COOH$	63
Octadecanoic acid	Stearic acid	$CH_3(CH_2)_{16}COOH$	70
Octadeca-9-enoic acid	Oleic acid	$CH_3(CH_2)_7CH{=}CH(CH_2)_7COOH$	13
12-Hydroxyoctadeca-9-enoic acid	Ricinoleic acid	$CH_3(CH_2)_5CH(OH)CH_2CH{=}CH(CH_2)_7COOH$	50
Octadeca-9-,12,15-trienoic acid	Linoleic acid	$CH_3CH_2CH{=}CHCH_2CH{=}CHCH_2CH{=}CH(CH_2)_7COOH$	−11
Docos-13-enoic acid	Erucic acid	$CH_3(CH_2)_7CH{=}CH(CH_2)_{11}COOH$	34

Table 3.14. *Common amino acids*

Systematic name	Common name	Structure
aminoacetic acid	glycine	NH_2CH_2COOH
2-aminopropanoic acid	alanine	$CH_3CH(NH_2)COOH$
3-aminopropanoic acid	β-alanine	$NH_2CH_2CH_2COOH$
2-amino-4-methyl-pentanoic acid	leucine	$(CH_3)_2CHCH_2CH(NH_2)COOH$
2-amino-3-phenyl propanoic acid	phenylalanine	$C_6H_5CH_2CH(NH_2)COOH$
2-amino pentanedioic acid	glutamic acid	$HOOCCH_2CH_2CH(NH_2)COOH$
2,6-aminohexanoic acid	lysine	$NH_2CH_2CH_2CH_2CH_2CH(NH_2)COOH$
3-mercapto 2-amino propanoic acid	cysteine	$HSCH_2CH(NH_2)COOH$
2-amino-4-(methylthio)butyric acid	methionine	$CH_3S(CH_2)_2CH(NH_2)COOH$

3.4.4 Carbohydrates

Carbohydrates are a large class of organic compounds consisting, exclusively, of carbon, hydrogen and oxygen atoms. They have the general formula $(CH_2O)_n$ but can be further subdivided as indicated below. Carbohydrates are formed by plants during photosynthesis.

$$nCO_2 + nH_2O \rightarrow (CH_2O)_n + nO_2 \qquad (3.9)$$

Monosaccharides

These are the simplest carbohydrates which contain three to seven carbon atoms and are not broken down by hydrolysis. The most common is glucose. It exists in a ring structure in equilibrium with a far lesser concentration of a linear aldehyde

$$(3.10)$$

Considering the ring structure, it may be seen that different compounds can exist with the same basic structure; the variations being dependent upon the position of an —OH group above or below the ring. Thus the structures of carbohydrates are drawn showing clearly the relative positions of functional groups, as below

glucose   mannose   fructose

Oligosaccharides and polysaccharides

Oligosaccharides are made up of two to five monosaccharide units, whilst polysaccharides consist of chains of larger numbers of units. Perhaps the best known disaccharide is sucrose, which is almost the sole component of domestic sugar:

sucrose

Sucrose is composed of glucose and fructose (a five carbon monosaccharide) and may be readily hydrolysed to the two constituent compounds.

Cellulose
is a glucose polymer and is the major structural material in plants. It is typically of molecular weight 300 000–500 000.

cellulose

Starch
is structurally very similar to cellulose, the only difference being in the bonding of the glucose units

starch

3.4.5 Humic substances

Decay of dead biological organisms in the environment gives rise to the formation of a range of substances of very varied and poorly defined chemical composition, known as humic substances. These compounds are present in soils, where they play an important role in determining the physical and chemical properties. As rain water passes through soils, it leaches humic substances which are consequently found also in rivers and sea water. The decomposition of marine organisms also contributes humic material to sea water. In both terrestrial and oceanic environments many such substances may play an important chemical role, particularly as complexing agents for trace metals, despite considerable differences in composition between humic substances from the two sources.

Traditionally, humic substances are divided into three groups according to their solubility:

(a) *fulvic acids*: soluble in acid and soluble in alkali;
(b) *humic acids*: insoluble in acid, but soluble in alkali;
(c) *humin*:   insoluble in both acid and alkali.

The chemical structure of humic substances has been the subject of much research, but is still not fully defined. It appears that they contain phenolic and carboxylic acid groups, which are responsible for their acidic properties, and aromatic rings. They are basically polymeric structures, built up of many aromatic sub-units. Fulvic acids have the lowest molecular weights, typically in the range 200–2000, whilst humic acids may be very large polymeric molecules with molecular weights ranging from a few thousand to several million.

The enormous significance of humic substances in environmental chemistry is now becoming fully understood, but much research is still needed to provide a fuller comprehension of these compounds.

3.5   Main block elements. Part 2: groups VB, VIA, VIIB, 0

3.5.1 Group VB, the nitrogen group

Nitrogen is an essential element and the most abundant in the earth's atmosphere where it is present primarily as the gas N_2. It is present in all amino acids and proteins (see Section 3.4.3).

Nitrogen forms a hydride, *ammonia*, NH_3, which is gaseous at environmental temperatures. It is highly water soluble, reacting to form *ammonium*, ions, NH_4^+ and hydroxide ions in a pH-dependent equilibrium:

$$NH_3 + H_2O \rightleftharpoons NH_4^+ + OH^-$$

Because of this ability to accept a hydrogen ion, ammonia is appreciably basic and is one of the few constituents of the natural environment capable of neutralising strong acids. The ammonium ion is a stable cation and ammonia forms salts such as ammonium nitrate, NH_4NO_3:

$$NH_3 + HNO_3 \rightleftharpoons NH_4NO_3$$

Although ammonium nitrate is a commercially produced solid used widely as a fertiliser, it has a marked tendency to revert to ammonia and nitric acid gases, a property which is especially important at the low concentrations found in the atmosphere. Ammonium chloride is also appreciably volatile, whilst ammonium sulphate is virtually involatile.

Nitrogen also forms a wide range of oxides. *Dinitrogen oxide*, N_2O, also known as nitrous oxide or laughing gas, is present in the lower atmosphere due to release from soils. It is of very low chemical reactivity and important only at the altitude of the stratosphere where it may be dissociated by solar ultraviolet light. *Nitrogen oxide*, NO, also known as nitric oxide, is a product of high temperature combustion processes. It is formed from combustion of nitrogenous materials in fuels, and from high temperature combination of molecular nitrogen and molecular oxygen:

$$N_2 + O_2 \rightarrow 2NO$$

NO is readily oxidised to *nitrogen dioxide*, NO_2, at high NO concentrations by molecular oxygen, and in the atmosphere by ozone, O_3. The chemistry of these processes is described in some detail in Section 2.1. Nitrogen dioxide has a tendency to form a dimer, *dinitrogen tetroxide*, N_2O_4:

$$2NO_2 \rightleftharpoons N_2O_4$$

Because of the low concentrations of NO_2 in the atmosphere, we know from the law of mass action that the compound exists as virtually free NO_2 molecules in this medium. A further oxide of nitrogen is the free radical species NO_3. This is of low stability, but despite a transient existence is important in atmospheric chemistry. It is formed primarily from reaction of nitrogen dioxide with ozone:

$$NO_2 + O_3 \rightarrow NO_3 + O_2$$

Reaction of this species with NO_2 leads to formation of *dinitrogen pentoxide*, N_2O_5:

$$NO_2 + NO_3 \rightleftharpoons N_2O_5$$

This is also a species of rather low stability with a transient existence in the atmosphere.

Nitrogen forms two environmentally important oxy-acids. These are *nitrous acid*, HNO_2, and *nitric acid*, HNO_3. The former is a weak acid, easily oxidised to the strong nitric acid. The corresponding anions *nitrite*, NO_2^-, and *nitrate*, NO_3^-, are common in the environment. In aquatic systems, nitrite is formed by biological oxidation of ammonium, and is converted subsequently to nitrate:

$$NH_4^+ \rightarrow NO_2^- \rightarrow NO_3^-$$

This process, known as *nitrification*, is important in converting ammonium, which is appreciably toxic to fish, to nitrate, an important plant nutrient which can contribute to eutrophication problems in rivers, lakes and estuaries.

Nitrates in water and food are also of concern because some studies indicate that under conditions prevailing in the human digestive system they can be converted to nitrosamines, which are compounds suspected of initiating cancer.

Phosphorus is an essential element for living organisms, and phosphates (PO_4^{3-}) are involved in electron transport mechanisms within living cells. Although anthropogenic input to the environment of phosphorus as phosphates is high, salt formation by reaction with Fe^{3+} and other metal ions can immobilise it.

Detergents containing polyphosphates are responsible for enrichment of natural waters in inorganic phosphates which contributes to eutrophication problems.

Arsenic is toxic to mammals, and environmental pollution originates from flue gas emissions during the production of iron, nickel, lead and cobalt and other metals. Geothermal waters can act as a natural source of arsenic to rivers and lakes. This type of source can have pronounced local effects, particularly where flux rates are enhanced due to the use of geothermal waters to generate electric power. In the last century it was used in pigments for household paints. Only in the late 1930s was it discovered that mould in walls could biomethylate arsenic to trimethyl-arsine, $(CH_3)_3As$, a foul smelling and highly toxic gas. It was the first case of biomethylation reported, and the mechanism of this process is believed to proceed via a series of additions of CH_3^+ groups after reduction of the oxidised species:

$$CH_3AsO_3^{2-} \xrightarrow{reduction} CH_3AsO_2^{2-} \xrightarrow{CH_3^+} (CH_3)_2AsO_2^- \xrightarrow{reduction}$$

$$(CH_3)_2AsO^- \xrightarrow{CH_3^+} (CH_3)_3AsO \xrightarrow{reduction} (CH_3)_3As$$

Although methyl arsenic compounds have been detected in the environment, low levels are not a cause for concern, except of course in the closed household environment. It is expected that antimony would be biomethylated in the environment, and there is an indication that stibinic $(CH_3)_2SbO(OH)$ and stibonic acids $CH_3SbO(OH)_2$ exist in the environment, but again at very low levels.

3.5.2 Group VIA, the oxygen group

Oxygen is the most abundant element in the earth's crust and a major constituent of the earth's atmosphere (21% by volume). By far the major portion of the oxygen in the atmosphere is in the form of diatomic molecules, O_2. A small proportion of elemental oxygen is also present in the form of ozone, O_3. Ozone is formed by photolysis of molecular oxygen in the stratosphere, followed by a combination of a resultant oxygen atom with an oxygen molecule in a collision involving an unreactive third body molecule, M:

$$O_2 + h\nu \rightarrow 2O \qquad (\lambda < 242 \text{ nm})$$

$$O + O_2 + M \rightarrow O_3$$

Ozone is a strong oxidising agent with an important role in atmospheric chemistry.

Oxygen is an essential element for life but O_2 is toxic at high concentrations in the mammalian bloodstream. It is the major constituent of water, and, because of its electronegativity, hydrogen bonding (Section 3.2.2) occurs in the bulk liquid. Water is a major solvent in the study of chemistry, and the complexing abilities of oxygen in it (H_2O acts as a ligand) determine the feasibility, reaction products and kinetics of a large number of reactions. Oxygen is present with hydrogen in organic compounds, which are the basis of life.

Oxidation reactions took their name from the ability of oxygen to acquire two electrons from other elements in bonding situations so that it can achieve an inert gas configuration. It is appreciated though that other elements have the ability to acquire electrons in a similar manner, hence oxidation was redefined in terms of electron transfer (Section 2.5.1). Oxygen plays an important role in atmospheric oxidation processes and in the oxidation (burning) of organic compounds. Dissolved oxygen in natural waters is essential to many forms of aquatic life. It is, however, consumed by organic materials in sewage and other effluents and hence treatment is necessary before discharge. The polluting capacity of effluents is described in terms of their biochemical oxygen demand (BOD) – their ability to consume dissolved oxygen, as measured at 20°C over five days in the laboratory.

Oxygen combines with many elements to form oxides which are common in the environment for both metals and non-metals. It is also an important component of many oxy-anions such as silicate (SiO_4^{4-}) or sulphate (SO_4^{2-}) which appear very commonly in the environment.

Sulphur is also an essential element and is contained in some amino acids and proteins. Like oxygen it may be found in an oxidation state of -2 in *hydrogen sulphide*, H_2S, but also exists in $+4$ and $+6$ states (e.g. SO_2, SO_4^{2-}, respectively).

Combustion of sulphur or of sulphur-containing fossil fuels leads to production of sulphur dioxide, SO_2, an important air pollutant:

$$S + O_2 \rightarrow SO_2$$

Strong oxidants or catalytic oxidation will convert sulphur dioxide to *sulphur trioxide*, SO_3:

$$2SO_2 + O_2 \rightarrow 2SO_3$$

Free sulphur trioxide is not known in the environment due to its rapid reaction with water.

Solution of SO_2 and SO_3 in water leads to formation of the corresponding oxy-acids sulphurous, H_2SO_3 and sulphuric, H_2SO_4, acids. Whilst the former is a weak acid, the latter is very strong (respective pK values are 6.9 and 1.9 for the dissociation of the second proton):

$$SO_2 + H_2O \rightarrow H_2SO_3 \text{ (sulphurous acid)}$$

$$SO_3 + H_2O \rightarrow H_2SO_4 \text{ (sulphuric acid)}$$

Organic sulphides encountered in the environment originate from biological degradation of sulphur-containing amino acids and proteins. For example, S-methylcysteine and methionine on degradation by certain bacteria would produce methyl mercaptan, CH_3SH, whilst methionine and S-methyl methionine would give rise to dimethyl sulphide, $(CH_3)_2S$. $(CH_3)S_2$ compounds are also known to occur naturally, whilst H_2S may also be the product of such biodegradation reactions. Naturally occurring sulphates (SO_4^{2-}) are also known to be methylated by certain microorganisms via successive methylations and reductions in a manner similar to the methylation of arsenic:

$$SO_4^{2-} \xrightarrow{\text{reduction}} SO_3^{2-} \xrightarrow{CH_3^+} CH_3SO_3^- \xrightarrow{\text{reduction}} CH_3SO_2^- \xrightarrow{CH_3^+}$$

$$(CH_3)_2SO_2 \xrightarrow{\text{reduction}} (CH_3)_2SO \xrightarrow{\text{reduction}} CH_3SH \xrightarrow{\text{dismutation}} (CH_3)_2S$$

The last step ($2CH_3SH \rightarrow (CH_3)_2S + H_2S$) is not a reaction unique to sulphur chemistry but S^{2-} ions in aqueous solutions have been observed to assist the dismutation of certain organometallic compounds:

$$2(CH_3)_3M^+ + S^{2-} \rightarrow$$
$$[(CH_3)_3M]_2S \rightarrow (CH_3)_4M + (CH_3)_2MS \quad (M = Sn, Pb)$$

Sulphide ions, HS^- and S^{2-}, are formed from sulphates by bacterial reduction in anoxic environments, and the presence of sulphide ions in aqueous environments or sediments is an indication of the anaerobic conditions that prevail in those environments. Because of the electronegativity of sulphur, S^{2-} is a good ligand for heavy metals, forming insoluble sulphides which are immobilised in anoxic sediments.

The chemistry of selenium is very similar to that of sulphur. It enters the atmosphere from combustion of fossil fuels. Methyl selenide compounds have been reported to be produced from action of fungi, and $(CH_3)_2Se_2$ can be produced from plants. However, selenium-containing amino acids do not exist.

3.5.3 Group VIIB, the halogens

The halogens are electronegative elements, with fluorine the most electronegative element of the periodic table. For this reason they form mainly ionic compounds in their -1 oxidation state. Fluoride is important in dental health, being essential to development of strong teeth. It is added to some municipal water supplies and is contained in most brands of toothpaste. However, excessive levels of dietary fluoride cause bone diseases and in some cases mottling of teeth. Hydrogen fluoride is a strong acid and will attack glass, SiO_2 and clay minerals. Fluoride emissions to the environment have increased in recent years from manufacture of polytetrafluoroethylene (PTFE) and other fluorinated plastics, and from aluminium smelting and production of phosphate fertilisers from fluorapatite rock:

$$CaF_2 \cdot 3Ca_3(PO_4)_2 + 10H_2SO_4 + 20H_2O \rightarrow 10CaSO_4 \cdot 2H_2O$$
(fluorapatite)

$$+ \ 2HF + 6H_3PO_4$$

The element chlorine is very abundant in sea water as chloride ions $(19.35 \ g \ kg^{-1})$. It is used in numerous industrial processes and released to the atmosphere due to fossil fuel burning and refuse incineration. It is also present as suspended sea salt aerosol in coastal regions. Gaseous hydrogen chloride can be released to the atmosphere from sea salt by reactions with less volatile acids, HNO_3 or H_2SO_4:

$$NaCl + HNO_3 \rightarrow HCl + NaNO_3$$

$$2NaCl + H_2SO_4 \rightarrow 2HCl + Na_2SO_4$$

Organochlorine compounds have found use as solvents (e.g. chloroform $CHCl_3$, carbon tetrachloride CCl_4), pesticides (e.g. DDT) and dielectric media (polychlorinated biphenyls (PCBs)), etc. It is also used in the manufacture of pvc (polyvinylchloride) and many other chemical products. Chlorine gas (Cl_2) is also used for disinfection of domestic water supplies.

In marine environments chlorine can be methylated to CH_3Cl which is volatile and plays a minor role in the global cycling of the element. Some of its organic compounds used as pesticides (e.g. DDT) are extremely toxic to mammals and are found to be persistent for long periods.

Both chlorine and fluorine are present in chlorofluorocarbon (CFC) compounds such as $CFCl_3$ and CF_2Cl_2. The effect of these compounds on stratospheric ozone is described in Section 5.1.5.

Bromine is commonly extracted from sea water and has a major use in the production of ethylene dibromide (EDB), which is a scavenger of lead in the motor fuel combustion process where alkylleads are used. EDB (systematic name 1,2-dibromoethane) is also used as a pesticide, but recently, after its toxicity to mammals was established, its use has diminished.

Iodine is an essential trace element and not a problem pollutant. Marine algae produce CH_3I, which enters the atmosphere, contributing to the global cycling of iodine.

3.5.4 Group 0, the noble gases

The noble gases are so named because until recently it was thought that they were too inert to undergo chemical reactions. This may still be the case for most of them, but a substantial chemistry of xenon has been established. They are not of particular environmental significance except for radon (Rn) which is radioactive. Radon is formed naturally in the decay chains of natural radioisotopes. Being a gas, once formed it diffuses slowly from rocks, soils and building materials, and can comprise a major source of exposure to radioactivity within the home.

Argon comprises 0.9% of dry air, but takes no part in atmospheric chemical reactions.

3.6   Transition elements

3.6.1 Periodic trends and general properties

Transition metals can be defined as elements having outer shell d orbitals involved in bonding or complex formation. In relation to the periodic table (Section 1.1.6) they show horizontal *and* vertical similarities in their physical and chemical properties. Horizontal similarities contrast with trends in s and p block elements where differences in traversing a row are striking. They are due to addition of inner d-electrons with relatively better shielding ability from increased nuclear charge in comparison to shielding ability of s and p electrons. The result is less marked differences in ionisation energies, electronegativities and atomic radii in traversing from scandium to zinc for example (Figures 3.5 and 3.6 and Table 1.2).

Some transition metals exhibit a large number of oxidation states (see Table 3.15) because a stable configuration is sought. For example, iron exhibits oxidation states of (II) as Fe^{2+} with an electron configuration of $[Ar]3d^6$ and (III) as Fe^{3+} with an electron configuration of $[Ar]3d^5$. In the former case two electrons from the 4s shell have been lost, whilst in the latter case two electrons from the 4s and one from the 3d shell have been lost, to achieve a stable d shell configuration (iron electronic configuration $[Ar]3d^64s^2$).

Values and variations in ionisation energies reveal trends in chemical properties of a series. There is a general increase in traversing a series and descending a group, due to an increase in effective nuclear charge. Deviations from the general trend occur because of difficulties in removing electrons from a filled (d^{10}) or half-filled (d^5) shell. Formation of both ionic and covalent compounds is contemplated because ionisation energies are greater than those of the -s block elements and lower than the p block elements. A large number of oxidation states for each

Figure 3.5. First and second ionisation energies of transition metals.

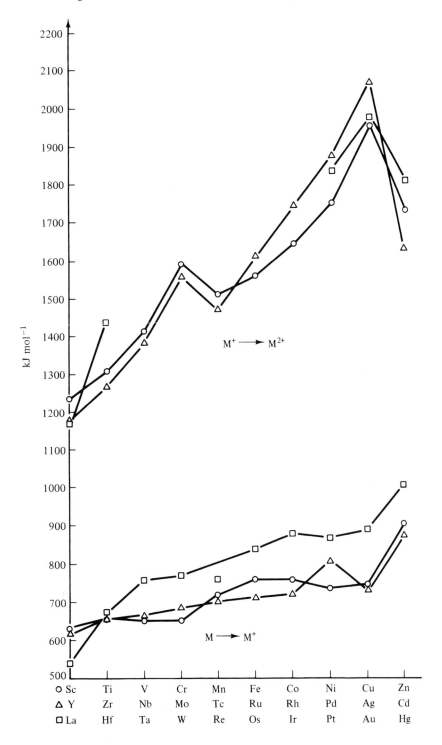

Table 3.15. *Known oxidation states of transition elements*

	I	II	III	IV	V	VI	VII	VIII
Sc			III	IV				
Ti		II	III	IV				
V	I	II	III	IV	V			
Cr	I	II	III	IV	V	VI		
Mn	I	II	III	IV	V	VI	VII	
Fe	I	II	III	IV	V	VI		
Co	I	II	III	IV	V			
Ni	I	II	III	IV				
Cu	I	II	III					
Y			III					
Zr			III	IV				
Nb	I	II	III	IV	V			
Mo	I	II	III	IV	V	VI		
Tc	I	II	III	IV	V	VI	VII	
Ru	I	II	III	IV	V	VI	VII	VIII
Rh	I	II	III	IV	V	VI		
Pd		II		IV				
Ag	I	II	III					
La			III					
Hf	I		III	IV				
Ta	I	II	III	IV	V			
W	I	II	III	IV	V	VI		
Re	I	II	III	IV	V	VI	VII	
Os	I	II	III	IV	V	VI	VII	VIII
Ir	I	II	III	IV	V	VI		
Pt		II		IV		VI		
Au	I	II	III					

Most common oxidation states are underlined

Figure 3.6. Atomic radii of transition metals.

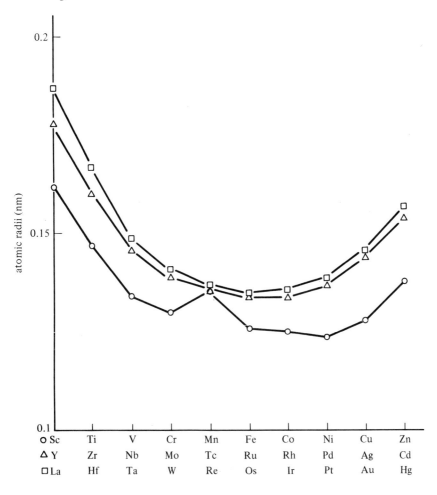

element can also be predicted because differences in successive ionisa-tion energies are small (Figure 3.5 and Table 3.15).

With increasing atomic number across a period, atomic radii decrease as a result of d orbital penetration of core electrons. Descending each group, atomic radii increase, but there is a smaller increase between fifth and sixth period than between fourth and fifth due to 4f electron penetration of the core electrons in the sixth period (lanthanide contrac-tion). Electronegativity increases as atomic number increases across a period but again differences are less marked (Table 1.2).

All elements are metals and good conductors of electricity. They may appear generally unreactive because of oxide film formation on their surfaces. They form alloys of variable composition because of their similar atomic sizes and find use as catalysts in a number of industrial processes. They form a large number of complexes because of their vacant d orbitals, the ability of effective nuclear charge to attract ligands at bonding distances and relatively small size due to d and f electron inner core penetration.

3.6.2 Group IIIA and the lanthanides

Scandium, yttrium and lanthanum form only compounds in the (III) oxidation state. There are no d electrons involved in complex formation or bonding, hence all compounds of this group resemble s-block element compounds in properties. Scandium behaves in a similar fashion to aluminium in reacting vigorously with water and forming hydrous oxides. Yttrium and lanthanum behave similarly but are more reactive; both show little tendency to form complexes. The lanthanides are fourteen elements between lanthanum and hafnium all having relatively similar properties to each other.

3.6.3 Group IVA: titanium, zirconium and hafnium

Titanium is a hard ductile metal with high strength to weight ratio and resistance to corrosion. Its chemical behaviour is similar to that of tin with Ti(IV) compounds being covalent and its salts, basic. The most stable oxidation state is (IV) with Ti(III) and Ti(II) formed by reaction of Ti(IV) with reducing agents. Ti(III) and Ti(II) compounds are reducing and Ti(II) compounds are stable in the solid state whilst aqueous solutions are strongly reducing (cf. $SnCl_2$). There is no aqueous Ti(II) chemistry but complexes can be made in non-aqueous media. Zirconium and hafnium have very similar chemistries because of similarities in ionisation energies, atomic radii, electronegativities, etc. In comparison with titanium they have the same main oxidation state of (IV) with very few known compounds of oxidation states of (III) and (II). Their oxides are more basic than that of titanium.

3.6.4 Group VA: vanadium, niobium and tantalum

The most stable oxidation states of vanadium are (II), (IV) and (V). The oxidation state (V) is mildly oxidising and the only stable compounds are oxides and fluorides. Mild reducing agents reduce V(V) to V(IV) which is more stable in aqueous solution than either V(V) or V(II). Molecular V(IV) oxides and tetrahalides are known except for VI_4. In aqueous solutions V(II) compounds are powerfully reducing and produced by reduction of higher oxidation states.

Unlike scandium and titanium, which have relatively little environmental significance, vanadium is essential to ascidians (sea squids and cucumbers, etc.) which concentrate it from sea water. It may have a beneficial effect in tooth decay and is thought to inhibit cholesterol biosynthesis in mammals, although it may be toxic to them if present in high concentrations in blood. It is emitted into the atmosphere from fuel oil combustion.

Niobium and tantalum have chemistries mainly in the V oxidation state with stable oxides and halides. Compounds of lower oxidation state can also be prepared by reduction of the higher oxidation states.

3.6.5 Group VIA: chromium, molybdenum and tungsten

Oxidation states of I to VI are known to occur for chromium but the most stable ones are III and VI. The pure metal is fairly unreactive and used in ornamental coatings. In acid solutions Cr(VI) is highly oxidising and the orange-coloured potassium dichromate, $K_2Cr_2O_7$, is used in volumetric titration analysis of reducing agents like I^- and Fe^{2+} in reactions of $I^- \rightarrow I_2$ and $Fe^{2+} \rightarrow Fe^{3+}$:

$$Cr_2O_7^{2-}(aq) + 14H^+(aq) + 6e^- \rightarrow 2Cr^{3+}(aq) + 7H_2O$$

Compounds and complexes of Cr(III) are invariably octahedral and stable in aqueous solutions, being oxidised only in strong alkaline conditions. Cr(VI) is highly toxic but Cr(III) less so. The former is an important toxic pollutant but little is known of its mode of action.

Molybdenum and tungsten have little known oxidation state III chemistries, and their VI oxidation state chemistry is not similar to the Cr(VI) one. They are both used in hardening steels, and tungsten in lamp filaments. Molybdenum complexes play an important role in fixation of atmospheric nitrogen.

3.6.6 Group VIIA: manganese, technetium and rhenium

Manganese has a stable oxidation state II chemistry because of the stability of the half-filled d^5 shell, and in this valence form it occurs in reducing environments (e.g. anoxic sediments). Polymeric MnO_2 is abundant in nature in ferromanganese nodules (see Section 5.5.2) and it has an oxidising surface due to non-stoichiometry (Section 1.3.2) and oxygen deficiency. Mn(II) forms an extensive series of salts with all common ions, most of which are soluble in water (except carbonates and phosphates). They are all stable in acidic but not alkaline solutions. One of the strongest oxidising agents is the permanganate ion ($Mn^{VII}O_4^-$). Widely used in volumetric analysis, it is hard to obtain pure, sensitive to light as an aqueous solution and slowly reduced by water to MnO_2. Its occurrence in the environment is most improbable.

Manganese is moderately toxic but essential to all organisms. Deficiencies are known to cause bone malfunctioning in chicks and infertility in mammals. It is also known to play a role in activating numerous enzymes.

Both technetium and rhenium are not known to have II oxidation state chemistries analogous to manganese but they have IV, V and VII state chemistries. Technetium is a synthetic element present as the isotope ^{99}Tc in effluent from nuclear fuel reprocessing facilities. It exists in aquatic systems substantially as the technetate ion, TcO_4^-.

3.6.7 Group VIII: Iron, ruthenium, osmium; cobalt, rhodium, iridium; nickel, palladium, platinum

Iron, because of its abundance and hardness, is widely used in modern civilisation. Its most stable oxidation states are II and III and the metal in moist air is moderately reactive forming hydrous ferric oxide $Fe_2O_3 \cdot nH_2O$ ('rust'). There is hardly any difference in stabilities of complexes or compounds of either oxidation state. Its complexation with oxygen and nitrogen ligands has been extensively studied because of its active biological chemistry. Iron III is a mild oxidising agent and its compounds are more covalent than the Fe(II) compounds.

Interconversion of Fe(II) and Fe(III) is rapid and accompanies changes in the environmental redox potential (Sections 2.5.4–2.5.6). In anoxic waters and sediments, iron is usually present as Fe(II), whereas oxic systems contain Fe(III). Hydrous oxides of Fe(III) are important as colloids in natural waters (Section 2.6), often in association with organic humic substances and trace metals such as lead. Manganese shows many similarities in behaviour to iron, with Mn(II) predominating in anoxic environments and Mn(VI) in oxic. Mn(IV) also forms hydrous oxides important in natural waters, although the oxidation of Mn(II) to Mn(IV) is far slower than for Fe(II) to Fe(III) and Mn(II) may persist for some time in oxic environments.

The chemistries of ruthenium and osmium are similar; neither is of especial environmental significance. Ruthenium is used in the production of catalyst systems for removal of oxides of nitrogen from vehicle exhaust gases.

Cobalt is a metal of low abundance in the earth's crust but widely distributed and biologically important. Vitamin B_{12} contains Co(III) in an octahedral configuration, and its coenzyme CH_3B_{12} has been shown to methylate a number of heavy metals including mercury and tin. Transfer of a CH_3^- group to other metals in the environment may make the latter more toxic than their inorganic counterparts and has resulted in extensive studies of the environmental methylation of heavy metals. The important oxidation states of cobalt are II and III, although Co(I) complexes are known to occur. Numerous oxo- hydroxo- and cyano-complexes have been identified, and octahedral polycyclic cobalt complexes have been synthesised to mimic and help understand the chemical behaviour of vitamin B_{12}.

Rhodium and iridium chemically behave in a similar manner to Co(III), but their environmental significance is limited. Nevertheless their complexes and chemistries are informative of metal–ligand complexing mechanisms.

Nickel can be found in nature in combination with sulphur, arsenic and antimony ores (e.g. pentlandite $((Ni, Fe)_9S_8$; millerite NiS) but is mined economically only in few areas of the world. Following periodic trends it exists at the low stable oxidation state of II with coordination numbers of 4, 5 or 6 with diverse and complex stereochemistries. The metal is relatively unreactive towards water, air or fluorides, hence it is used, by electroplating, for protection of other metals. It is relatively toxic to most plants but much less so to mammals, and it may pollute areas locally where it is mined or worked.

Palladium and platinum mainly exist in the oxidation states of II and IV and their chemistries bear some similarities to that of nickel. They are the by-products of nickel mining, relatively unreactive as metals. Both are used as oxidation catalysts to remove unburned hydrocarbons and carbon monoxide from vehicle exhaust.

3.6.8 Group IB: copper, silver, gold

Copper mainly occurs in the oxidation states I and II and is an element known and used from ancient times (Bronze Age). Its minerals and compounds are widely abundant on the earth's crust as sulphides, arsenides, chlorides and carbonates, and it is still used in alloys such as brass (Cu–Zn) and bronze (Cu–Sn). The metal is oxidised in moist air forming a green coating of carbonate. Its thermal and electrical conductivities are well known and utilised domestically and industrially. Copper is a constituent of O_2-transporting pigments and redox enzymes, hence essential to life. However, in large quantities it is toxic to plants and invertebrates but less so to mammals. Pollution by copper is centred upon areas of industrial use and agricultural applications.

Silver and gold occur mainly as sulphides and arsenides in nature, although silver chlorides also occur. Their chemistries differ from each other and from that of copper. Silver mainly exists in the I oxidation state, having extremely high affinity for halides, forming water insoluble compounds. They are relatively unreactive elements forming numerous alloys, useful in industrial application because of their inertness.

3.7   Biologically essential elements

There are some chemical elements which are essential to the survival of biological organisms. These are not necessarily the same elements for each organism, although generally few differences appear to exist; this is not reliably known since it is difficult to establish exactly which elements are essential and which are not. Table 3.16 shows some elements known to be essential in humans.

Essential elements are usually divided into two categories according to the amounts in which they are required. *Macronutrients* are those required in appreciable quantities and which take a major part in bodily structure or function (e.g. nitrogen as a component of amino acids and proteins). Elements required only in trace amounts are termed *micronutrients*. These play a lesser, but nonetheless essential, role by, for example, taking part in enzymic reaction processes.

For essential elements, the body has a homeostatic, or self-regulating, mechanism to balance intake and excretion and hence control bodily levels. This mechanism may not be able to cope with very high intakes of the element, in which instance toxic effects may appear. An example is in the exposure of industrial workers to high levels of zinc fume (fine particles of zinc oxide) which are very effectively absorbed through the lungs. The resultant medical condition is known as 'zinc fume fever'.

A picture thus emerges for essential elements exemplified by Figure 3.7. At inadequate levels of intake, deleterious effects arise, known as a deficiency syndrome. Above the acceptable exposure window, toxic effects may be observed.

In the case of non-essential elements, the body has no need for the element at all. Generally it can, however, cope with modest exposures

Table 3.16. *Essential elements and their functions*

Element	Function
Macronutrients	
Nitrogen (N)	Structural component of amino acids, proteins, enzymes, hormones etc.
Phosphorus (P)	Structural components of ATP, nucleic acids, phospholipids etc., involved in electron transfer processes (energy transfer)
Sulphur (S)	Structural component of some amino acids, proteins and vitamins
Potassium (K)	Important in sustaining ionic balance inside and outside cells, directly in carbohydrate metabolism as cofactor of enzymes in protein
Sodium (Na)	Important for ionic balance of cells
Magnesium (Mg)	Similar functions to potassium
Calcium (Ca)	Activator of several enzymes, important in permeability of membranes necessary for cell wall formation
Iron (Fe)	Component of iron-porphyrins (haems)
Micronutrients	
Boron (B)	Essential but exact function unknown. It may play role in translocation of sugars
Manganese (Mn)	Involved in nitrogen metabolism, cellular respiration and acts as cofactor in activating numerous enzymes
Chlorine (Cl)	A major anion
Zinc (Zn)	Component of metalloenzymes, activator of enzymes; important role in metabolism of proteins and nucleic acids
Copper (Cu)	Structural component of enzymes that catalyse oxidation reactions
Iodine (I)	Constituent of thyroid hormones which regulates growth in vertebrates
Cobalt (Co)	Constituent of vitamin B_{12}; required for nitrogen fixation
Molybdenum (Mo)	Structural component of enzyme reducing nitrate to nitrite; essential for nitrogen fixation
Fluorine (F)	Essentiality has not been proven for any organism but helps prevent teeth decay

Figure 3.7. Exposure in relation to requirement for an essential element.

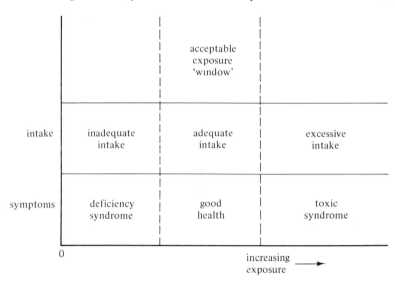

without deleterious effect, although elevated levels of exposure again give rise to toxic effects.

3.8   Organometallic compounds

Only compounds with a metal to carbon bond can be classified as organometallic. This definition usually excludes carbides containing for example the C_2^{2-} ion, and sometimes carbonyls (metal atoms coordinated by CO molecules). Many organometallics resemble organic compounds rather than inorganic. They are important environmentally because they are less polar than their inorganic counterparts and more soluble in lipids. By penetrating cell membranes they can interfere with electron transfer mechanisms inside the cell. Their toxicity arises mainly from such behaviour.

Organometallics can be classified according to their type of bond with carbon (Figure 3.8). The metal to carbon strength decreases as groups are descended because overlap of carbon (sp^3) orbitals decreases as heavier metals have more diffuse orbitals. Stability of organometallics should be considered in terms of thermodynamic stability of the bond, stability to oxidation and with respect to hydrolysis. Thermodynamic stability is directly related to the bond strength (Table 3.17). All organometallics are unstable with respect to oxidation, the driving force being the large negative ΔG for formation of CO_2, H_2O and metal

Figure 3.8. Organometallic compounds classified according to bond type.

Li	Be	electron deficient structures										B	C	N
		e.g. $[(CH_3)_3 Al]_2$												
Na	Mg											Al	Si	P
K	Ca	Sc	Ti	V	Cr	Mn	Fe	Co	Ni	Cu	Zn	Ga	Ge	As
Rb	Sr	Y	Zr	Nb	Mo	Tc	Ru	Rh	Pd	Ag	Cd	In	Sn	Sb
Cs	Ba	La	Hf	Ta	W	Re	Os	Ir	Pt	Au	Hg	Tl	Pb	Bi
Fr	Ra	Ac												

ionic
metal
to carbon
bonds;
e.g. $(C_2 H_5)$ Na

forming both σ and π metal
to carbon bonds

Forming σ metal to
carbon bonds;
covalent, volatile
compounds

oxide. Many are also kinetically unstable (i.e. activation energies for reactions are low; see Chapter 2) because of lone pairs of electrons or empty low-lying orbitals on the metal which are available for coordination of oxygen molecules e.g. $(CH_3)_3In: \leftarrow O_2$. For similar reasons hydrolysis can be facilitated, e.g.

$$R_2M + 2H_2O \rightarrow 2RH + M(OH)_2$$

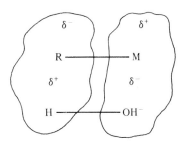

Table 3.17. *Some mean metal to carbon dissociation energies for main group methyl compounds* $(kJ\ mol^{-1})$

$(CH_3)_2Zn$	176	$(CH_3)_4Pb$	155
$(CH_3)_2Cd$	138	$(CH_3)_3As$	250
$(CH_3)_2Hg$	121	$(CH_3)_3Sb$	217
$(CH_3)_4Sn$	217	$(CH_3)_3Bi$	142

The rate of hydrolysis depends upon the polarity of the M—C bond, hence relatively ionic organometallic bonds (Figure 3.8) would be more susceptible to hydrolysis than covalent ones (e.g. $(CH_3)_4Pb$ is not susceptible to hydrolysis).

Metal to carbon bonds are formed by the action of

(a) organic halides:

$$nRX + M \rightarrow R_nMX_n \text{ or } R_nM + MX_n$$
$$\text{ether}$$

e.g.
$$RX + Mg \rightarrow RMgX$$

$$2Al + 3RX \rightarrow R_3Al_2X_3$$

$$4PbNa + 4C_2H_5Cl \rightarrow (C_2H_5)_4Pb + 4NaCl + 3Pb$$

and

(b) metal exchange reactions:

e.g.
$$Zn + (CH_3)_2Hg \rightarrow (CH_3)_2Zn + Hg$$

$$2RMgX + CdX_2 \rightarrow R_2Cd + 2MgX_2$$

In the environment, variations of these reactions may take place inside the living cell with alkyls originating from natural methylating agents (CH_3CoB_{12}, methionine, etc.) or may take place in aqueous solutions by methylating metabolites (e.g. CH_3Cl, $(CH_3)_2S$, etc.). Environmentally important organometallic compounds have been considered under the appropriate group of the periodic table.

3.9   Radionuclides

An atom whose nucleus contains a specified number of protons and neutrons is termed a *nuclide*. Nuclei with an unfavourable proton-/neutron ratio will undergo a nuclear disintegration to achieve a more stable configuration. As this process is accompanied by the emission of radiation, such an unstable nuclide is known as a *radionuclide*. The decay process proceeds at a well-defined rate characterised by the radionuclide considered. This attribute may be exploited to date materials, both geological and biological in nature, and to determine the kinetics of environmental processes such as water mixing and sediment deposition. Alternatively, artificial radionuclides and natural radionuclides with enhanced concentrations due to anthropogenic influences may be used as tracers for water masses. Some radionuclides deserve special consideration due to the threat they may pose as environmental pollutants.

3.9.1 Radioactive decay

Several modes of nuclear disintegration exist by which an unstable nucleus may achieve stability. The particular nuclear reaction which proceeds depends upon the nature of the instability. *Beta decay* occurs when a nucleus contains an excess of neutrons. A neutron is transformed into a proton with the consequent emission of an energised electron known as a *negative beta (β^-) particle*:

$$n \rightarrow p^+ + \beta^-$$

As the mass of an electron is negligible, there is only a minor change in atomic mass but the atomic number increases by one. A nucleus with a proton excess will tend to decrease its positive charge and hence atomic number. Two mechanisms are possible. Firstly, this may occur by *positron decay*, in which a positively charged electron (β^+ or e^+) is emitted when a proton is transformed into a neutron:

$$p^+ \rightarrow n + e^+$$

Alternatively, stability may be enhanced by *electron capture* (EC). In this instance the nucleus attracts an inner orbital (K or possibly L shell) electron which reacts with a proton to create a neutron:

$$p^+ + e^- \rightarrow n$$

The resulting daughter element exists in an excited electronic state. Electronic rearrangement occurs in order to fill the vacancy in an inner orbital and consequently X-ray radiation is emitted.

Heavy unstable nuclides may undergo *alpha decay*. Energy is lost by the emission of an alpha (α) particle, which is a helium nucleus, that is it has two protons and two neutrons. Alternatively, *fission* may occur, in which case a large nucleus splits to produce lighter daughter elements.

A nucleus may exist in an excited state in a fashion analogous to an electronic excited state. The re-establishment of ground state conditions results in the emission of high frequency energy. This process is termed *isometric transition* and the radiation is in the γ-ray spectrum. Gamma emission often accompanies the ejection of particles (β^-, e^+ and α) from the nucleus. Radioactive decay processes are summarised together with examples in Table 3.18.

Worked example

What are the daughter products following the radioactive decay of (i) ^3_1H, (ii) $^{26}_{13}Al$ and (iii) $^{239}_{94}Pu$. Refer to Table 3.19 for emission modes.

(i) ^3_1H disintegration modes: β^- emission

$$^3_1H \xrightarrow{\beta^-} {}^3_2He$$

(ii) $^{26}_{13}Al$ disintegration modes: β^+ emission and electron capture

$$^{26}_{13}Al \xrightarrow{\beta^+} {}^{26}_{12}Mg$$

$$^{26}_{13}Al \xrightarrow{EC} {}^{26}_{12}Mg$$

(iii) $^{239}_{94}Pu$ disintegration modes: α emission

$$^{239}_{94}Pu \xrightarrow{\alpha} {}^{235}_{92}U$$

3.9.2 Dating techniques

The distintegration rate of a radionuclide depends solely upon the structure of the nucleus and is independent of the chemical and physical state of the atom. In kinetic terms radioactive decay is a first order process having a radionuclide-specific rate characterised by the *decay constant*, λ. This may be expressed as:

$$\frac{dN}{dt} = -\lambda N \tag{3.11}$$

where N is the number of radioactive atoms and dN/dt is the rate of nuclear disintegration, also known as *activity*. The SI unit of activity is the becquerel (Bq) which is one disintegration per second. The historic unit is the curie (Ci), which equals 3.700×10^{10} Bq.

Table 3.18. *Modes of nuclear disintegration*

Disintegration mode	Effect on nuclide		Example
	charge Z	mass A	
Beta (β^-) emission	+1	0	$^{14}_6C \rightarrow {}^{14}_7N + \beta^-$
Positron (e^+) emission	−1	0	$^{22}_{11}Na \rightarrow {}^{22}_{10}Ne + e^+$
Electron capture	−1	0	$^7_4Be + e^- \rightarrow {}^7_3Li$
Alpha (α) emission	−2	−4	$^{238}_{92}U \rightarrow {}^{234}_{90}Th + {}^4_2He^{2+}$
Fission			$^{252}_{98}Cf \rightarrow$ several products
Isometric transition	0	0	$^{60m}_{27}Co \rightarrow {}^{60}_{27}Co$

Integrating the fundamental decay equation (3.11) using the initial conditions that $N = N_0$ at $t = 0$ gives:

$$N = N_0 e^{-\lambda t} \qquad (3.12)$$

or

$$\ln \frac{N_0}{N} = \lambda t \qquad (3.13)$$

If we consider the case when one half of the radionuclides originally present have decayed, then $N = N_0/2$. The time necessary for this to happen, known as the *half life*, $\tau_{\frac{1}{2}}$, would then be:

$$\tau_{\frac{1}{2}} = \frac{\ln 2}{\lambda} = \frac{0.693}{\lambda} \qquad (3.14)$$

Table 3.19. *Disintegration modes and half lives of some radionuclides*

Radionuclide	Disintegration mode	Half life
$^{3}_{1}H$	β^-	12.26 y
$^{7}_{4}Be$	EC	53.37 d
$^{10}_{4}Be$	β^-	2.5×10^6 y
$^{14}_{6}C$	β^-	5730 y
$^{22}_{11}Na$	β^+, EC	2.602 y
$^{26}_{13}Al$	β^+, EC	7.4×10^5 y
$^{35}_{16}S$	β^-	88 d
$^{40}_{19}K$	β^-, β^+, EC	1.28×10^9 y
$^{65}_{30}Zn$	β^+, EC	243.6 d
$^{87}_{37}Rb$	β^-	5×10^{11} y
$^{90}_{38}Sr$	β^-	28.1 y
$^{103}_{44}Ru$	β^-	39.6 d
$^{131}_{53}I$	β^-	8.070 d
$^{137}_{55}Cs$	β^-	30.23 y
$^{210}_{82}Pb$	β^-, α	21 y
$^{222}_{86}Rn$	α	3.823 d
$^{235}_{92}U$	α, SF	7.1×10^8 y
$^{238}_{92}U$	α, SF	4.51×10^9 y
$^{239}_{94}Pu$	α	2.44×10^4 y

EC = electron capture; SF = spontaneous fission.

The half lives of several radionuclides are given in Table 3.19. The great variation in decay rates allows a wide choice of time scales for dating purposes, ranging from millions of years necessary for dating rocks, to tens of years as required for determining coastal sedimentation rates. Two criteria must be satisfied in order that radioactive dating techniques may be applied. Firstly, a closed system is necessary to ensure that no loss of daughter nor parent nuclides occurs. Secondly, the initial number of daughter nuclides relative to parent nuclides must be known, although in several instances this initial number may be assumed to be zero. The following examples outline the principles of radioactive dating.

Potassium-40

Potassium-40 has a half life of 1.26×10^9 years and may be utilised to determine the age (in millions of years) of volcanic material. Radiogenic products of ^{40}K are ^{40}Ca (89%, β^- decay) and ^{40}Ar (11%, electron capture). These isotopes exhibit a greatest natural abundance for both elements. Calcium-40 is inapplicable for geochronological purposes due to its relatively high concentration in most rock material and therefore ^{40}Ar is utilised. Thus, potassium/argon dating methods must assume that no atmospheric ^{40}Ar is incorporated into the rock matrix at the time of formation and that radiogenic ^{40}Ar subsequently produced cannot diffuse out of the rock. The geochronology of the rock is determined from the measured concentrations of the ^{40}K (parent isotope) and the ^{40}Ar (daughter product). As noted above, only a fraction of the ^{40}K actually decays to ^{40}Ar and so that the amount of this daughter product must be scaled upward to include that parent material which would have formed ^{40}Ca.

Assuming a closed system ensures that at time t the number (D) of stable radiogenic daughter atoms present plus the number (N) of remaining radionuclides equals the number (N_0) initially present. As N_0 equals $N + D$, equation (3.13) may be expressed as:

$$t = \frac{1}{\lambda} \ln \left(\frac{N + D}{N} \right) \tag{3.15}$$

As noted above, the daughter product quantified is only one possible decay product and so must be corrected. For dating purposes equation (3.15) becomes

$$t = \frac{1}{\lambda} \ln \left\{ 1 + \frac{D}{N} \left(1 + \frac{1}{R} \right) \right\}$$

where R is the efficiency of disintegration via electron capture (^{40}Ar) relative to β^- emission (^{40}Ca); in this case $R = 11/89$.

Carbon-14

Carbon-14 has a half life of 5730 years and can be used to date biological materials. Carbon-14 is produced in the atmosphere from the collision of neutrons with nitrogen:

$$^{14}_{7}N + ^{1}_{0}n \rightarrow ^{14}_{6}C + ^{1}_{1}H$$

This ^{14}C becomes incorporated firstly into atmospheric carbon dioxide and then in turn into the living tissue of plants and animals. The application of radiocarbon dating techniques requires the following basic assumptions:

(i) living tissue achieves a steady state with respect to atmospheric ^{14}C;

(ii) the specific activity (i.e. the ratio of total activity of a radio-nuclide to the total mass of that element present) of atmospheric ^{14}C is invariant with time, approximately 918 Bq g^{-1} of carbon. The recent increase in the relative abundance of ^{14}C in the atmosphere results from nuclear weapons testing in the 1950s and 1960s.

Upon the death of an organism the steady state condition cannot be maintained and the concentration of ^{14}C decreases.

Lead-210

Lead-210 has a half life of 21 years, which makes it ideal for dating recently deposited sediments in lakes, estuaries and coastal marine environments. The ^{210}Pb results from the decay chain of ^{226}Ra (Table 3.20). The main principle of using ^{210}Pb rests with the first daughter product in this decay chain. ^{226}Ra decays slowly to the gaseous radionuclide ^{222}Rn which is emitted to the atmosphere from sediments, aerosols or rocks. This gas is dispersed within the atmosphere and decays quickly to ^{210}Pb. A relatively uniform ubiquitous fallout of this long-lived product occurs. Transfer from the atmosphere to the litho-

Table 3.20. *The decay chain of* ^{226}Ra

Radionuclide	Decay process	Half life
^{226}Ra	α	1620 y
^{222}Rn	α	3.8 d
^{218}Po	α	3.05 min
^{214}Pb	β^-	26.8 min
^{214}Bi	β^-	19.7 min
^{214}Po	α	1.6×10^{-4} s
^{210}Pb	β^-	21 y
^{210}Bi	β^-	5.0 d
^{210}Po	α	138 d
^{206}Pb	stable	

sphere takes about a month. Profiles of ^{210}Pb with depth in a sediment show a decrease in activity due to the decay chain to the stable ^{206}Pb. The amount of ^{210}Pb at any depth allows the age to be estimated. Several assumptions must be made:

(i) there is a uniform flux of ^{210}Pb to the sediments;
(ii) the sedimentation rate itself is constant;
(iii) the sediments are not disturbed upon deposition (i.e. bioturbation, slumping);
(iv) ^{210}Pb is not affected by diagenetic processes (i.e. chemical remobilisation following the deposition of the sediment – see Section 5.5.4);
(v) an undisturbed core can be collected.

It should be noted that the presence of ^{226}Ra in the sediment itself will lead to the *in situ* formation of ^{210}Pb. Hence, the total ^{210}Pb present must be corrected to account for this material.

Worked example
What is the age of a sediment which exhibits a ^{14}C activity of 400 Bq g^{-1}?

$$\tau_{\frac{1}{2}} = 5730 \text{ y}$$

$$\lambda = \frac{0.693}{5730} = 1.21 \times 10^{-4} \text{ y}^{-1}$$

$$t = \frac{1}{\lambda} \ln \frac{N_0}{N}$$

$$t = \frac{1}{1.21 \times 10^{-4}} \ln \frac{918}{400}$$

$$t = 6870 \text{ y}$$

3.9.3 Radionuclides of anthropogenic origin

Man's activities have given rise to a suite of isotopes which might not otherwise have been expected to be present. These are often referred to as artificial radioisotopes. Also, concentrations of natural isotopes have been enhanced well above normal levels for many radioisotopes. These enrichments result from nuclear processes, both power generation and weapons testing. It should be pointed out that analytical techniques for the determination of several radioisotopes are extremely sensitive. Thus, even with large dilution effects, the radioisotopes can be detected at great distances from their point of origin. Such analytical attributes have contributed to the utility of radioisotopes as tracers, even on a global scale, and perhaps to public misconceptions as to their risk.

Atmospheric testing of nuclear weapons, particularly prominent in the 1950s and early 1960s, gave rise to such natural isotopes as tritium (^{3}H) and ^{14}C. The testing caused large increases over the previous background concentrations. Furthermore, these constituents were globally dispersed. These features together resulted in a marked enhancement in the flux of ^{3}H and ^{14}C to the surface waters of the ocean. In polar latitudes, downwelling of waters forms deep ocean waters which circulate around the world at quite slow rates. The relatively sudden injection of artificially enhanced concentrations of the radioisotopes has enabled the movement of such waters to be traced and current velocities estimated.

Nuclear power generation has also been accompanied by the release, both knowingly and accidentally, of radioisotopes. Permitted releases have most often resulted from nuclear fuels reprocessing. The reprocessing plant at Sellafield in Cumbria, north England, has long discharged effluent into the Irish Sea. The emissions contain a cocktail of isotopes, including several transuranic elements. Plutonium has been observed in the nearby village of Ravenglass, following a complicated pathway of sea to land transfer. However, as with weapons testing, some benefits have been gleaned. The discharge contains both ^{134}Cs and ^{137}Cs. Being the same element, these isotopes would be expected to

Figure 3.9. Gamma spectra of the IMER building air filter, sampled on May 2, 1986, illustrating the most prominent peaks. Analyses by Ge(Li) detector. Potassium-40 is from the natural background. (Compiled from Hamilton, Zou & Clifton, 1986.)

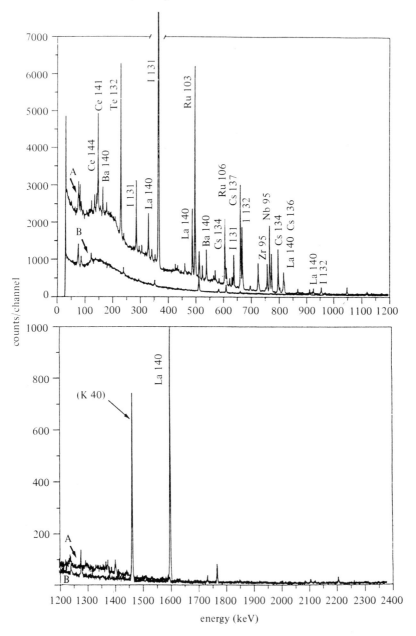

behave similarly in the marine environment. The isotopes do have different half lives (2.2 years for ^{134}Cs and 28 years for ^{137}Cs). Therefore, differences in the ratio of ^{134}Cs:^{137}Cs can be used to estimate the length of time that has elapsed since discharge, thereby allowing the determination of current velocities in the Irish Sea or sedimentation rates in adjacent estuaries. Not all discharges from Sellafield have been planned, and permissible levels of radioisotope emissions to the sea have been accidentally exceeded, leading to closure of adjacent beaches on at least one occasion. Sellafield also emits radioisotopes, such as ^{137}Cs and ^{90}Sr, to the atmosphere. Relatively large particles (10 to 100 μm diameter) of uranium oxide were released in the period 1954–7. Such particles would have only localised influences. The associated isotopes included ^{90}Sr, ^{137}Cs and $^{239+240}$Pu.

Nuclear power plants themselves have also acted as sources of contamination. While global dispersion of small amounts of plutonium has been caused by the disintegration of nuclear powered satellites upon re-entry into the atmosphere, the best known example of an accidental discharge is that due to the Chernobyl nuclear power plant in the Ukraine. Substantial quantities of radioactive pollutants were released following an accident at this plant on April 26, 1986. Several isotopes were emitted, as illustrated in Figure 3.9, the gamma spectrum of an air filter collected in southern England during the week ending May 2, 1986. This incident resulted in the global dispersion of several radio-isotopes, but the fallout, especially that mediated by rainfall, has caused appreciable contamination in some regions of Europe with subsequent restrictions on the slaughter of lamb and reindeer for human consumption.

Questions

(1) The main group elements are also called s- and p-block elements. Explain why and list their subdivisions in groups.

(2) Define atomic radius, ionisation energy and electronegativity. Comment on their periodic trends.

(3) Why are the group IA elements called alkali metals? Which compounds contribute to water hardness and where do they originate from?

(4) Suggest routes by which group IIB elements enter the environment and define the terms 'biomethylation' and 'bioamplification'.

(5) The group IVB organometallic compounds are considered highly toxic. Elaborate on their industrial production, their usage and their environmental transformations.

(6) Discuss the environmental impact of the group VB elements and their compounds.

(7) How does sulphur play a role in the 'acid rain' problem and how does it influence other environmental processes (see also Chapter 5)?

(8) Indicate which elements are considered essential and discuss the concepts of 'deficiency syndrome' and 'acceptable exposure window'.

(9) Which compounds are classified as organometallic? Explain the term 'stability of organometallic compounds'.

(10) Describe the modes of nuclear disintegration with examples.

(11) Define the terms 'decay constant', 'activity' and 'half life'.

(12) Explain the basis of carbon-14 as a dating tool. What criteria must it fulfil to be used successfully?

(13) If milk powder contaminated by radioactive fallout has an initial activity of 100 Bq kg^{-1} of ^{90}Sr and 100 Bq kg^{-1} of ^{131}I, what are the respective activities if the powder is stored for one year?

(14) Write down all the possible structural isomers of heptane.

(15) Write down all the possible geometric and structural isomers of hexene.

(16) A molecule has the composition C_8H_{10}. Write down structures of aliphatic and aromatic hydrocarbons having this molecular formula.

(17) Explain how both amino acids and monosaccharides can join together to form chain-like structures. How does cellulose differ from starch in chemical structure?

(18) Explain how fatty acids and glycerol are obtained from fats, indicating the chemical structure of glycerol and of the more common fatty acids.

(19) Write down general formulae for the four types of amine, giving an example of each.

Reference

Hamilton, E.I., Zou, B. & Clifton, R.J. (1986). *Sci. Total Environ.* **57**, 231–51.

Further reading

Cox, P.A. (1989). *The Elements*: *Their Origin, Abundance and Distribution*. Oxford University Press.

Dasent, W.E. (1982). *Inorganic Energetics*: *an introduction*, 2nd edn. Cambridge University Press.

Fergusson, J.E. (1982). *Inorganic Chemistry and the Earth*. Oxford: Pergamon Press.

Morrison, R.T. & Boyd, R.N. (1987). *Organic Chemistry*, 5th edn. Boston: Allyn and Bacon.

Weast, R.C. (1979). *CRC Handbook of Chemistry and Physics*, 59th edn. West Palm Beach: CRC Press, Inc.

4

○ ○ ○ ○ ○ ○ ○ ○ ○ ○ ○ ○ ○ ○ ○ ○ ○ ○ ○ ○

Analytical chemistry

4.1 General principles of analytical chemistry

4.1.1 Introduction

Analytical chemistry encompasses the science of characterising a sample. Two broad categories can be considered. *Qualitative analysis* deals with the question of what elements or compounds are present. In the simplest case, this may involve only sample identification. For instance, is an unknown water sample fresh or saline? Several screening tests may rely on determining whether or not a particular substance is present or absent. This has applications in quality control during food processing where the analysis might serve to distinguish foreign bodies. Blood samples from athletes and racehorses are routinely examined for the presence of prohibited drugs. Qualitative analyses may also be used to trace the source of a sample. Pollen analysis of honey indicates the country of origin. Comparing the 'fingerprint' of substances in an oil slick with that in the bilge tanks of ships may enable the responsible vessel to be identified. *Quantitative analysis* involves determining how much of an element or molecule is present. Several types of quantitative analyses can be contemplated. These are exemplified below using the analysis of petrol (gasoline). A *complete* analysis would quantify the amount of each constituent in the sample. This would involve the determination of hexane, octane, toluene, tetraethyllead, etc. Such analyses are generally impossible and for environmental samples usually unnecessary. For an *ultimate* analysis the amount of each element is quantified. The composition of petrol would therefore be expressed in terms of carbon, hydrogen, oxygen, sulphur, lead, etc. A *partial* analysis involves measuring the concentration of only certain constituents, such as, for instance, toluene or tetramethyllead. Partial analyses would be the most utilised type of analysis in environmental chemistry. They would have applications in ensuring that statutory limitations are obeyed. This could involve monitoring the chlorine content of drinking water leaving a water treatment works, or the chromium content in the effluent from an electroplating plant. Finally, a *speciation* analysis assesses the amount of each form in which an element may exist in a sample. Considering the speciation of lead in the exhaust gases of a vehicle, one would have to measure the concentrations of $(CH_3)_4Pb$, $(C_2H_5)_4Pb$, $PbBrCl$, $PbBr_2$, etc.

The role of the analyst involves the planning and execution of the desired analysis. Firstly, he must define the problem. In analytical terms, this means establishing what parameters must be measured. The problem must be defined in collaboration with those asking the question. If someone asks an analyst 'Is this water safe to drink?', would that person be satisfied to know that the lead concentration does not exceed the accepted limit of $50\ \mu g\ dm^{-3}$? The second task for the analyst is to choose the best technique for the analysis. This can involve the development of a novel procedure or the improvement of existing instrumentation. Thus, the analyst must be familiar with the chemistry involved in the procedure and principles of the instrumentation. An understanding of the relationship between the analyte concentration and the measurement procedure ensures that anomalous data can be evaluated and analyses repeated possibly using a different methodology.

Several criteria influence the selection of the technique for any particular analysis. These are:

  (i) accuracy;
  (ii) precision;
  (iii) sensitivity and detection limit;
  (iv) amount of sample;
  (v) selectivity or specificity;
  (vi) nature of the sample;
  (vii) destructive or non-destructive techniques;
  (viii) reliability;
  (ix) simplicity;
  (x) batch or continuous analysis;
  (xi) time for analysis;
  (xii) cost.

Many of these criteria are self-evident. *Accuracy* refers to the degree of agreement between a measured value and the true value. Often the absolute true value is not known and accuracy is practically defined in terms of an 'accepted' true value. This accepted true value is best obtained by having the analyte quantified using several techniques and preferably by different laboratories. International standards of plant and animal tissue, sea water, sediments, fuels, etc., can be purchased. These tend to be expensive as several elements have certified concentrations, the result of many analyses in a number of institutes.

Precision is the degree of agreement between replicate measurements of the same quantity. This then indicates the reproducibility of a result and is best denoted by recording the standard deviation of a set of measurements. This is considered further when discussing errors. It should be noted here that good precision does not guarantee good accuracy. Considering the results of two analysts depicted in Table 4.1, analyst 1 displays better precision but analyst 2 is more accurate.

Sensitivity is the response of the instrument per unit weight of the determinant, whilst *detection limit* expresses the minimum concentration or amount of analyte which can be confidently discriminated from zero. A technique must be employed that is sufficiently sensitive to quantify the analyte. However, the absolute amount of analyte presented can have an influence in that the material could be preconcentrated prior to the analysis. The *selectivity* refers to the preference exhibited by a reaction or test toward the substance of interest over other (interfering) substances. Ideally the analyst would choose a *specific* reaction or test for which only the analyte responds.

The nature of a sample will affect the choice of analytical technique.

Table 4.1. *Accuracy and precision exhibited by a pair of analysts measuring the same analyte in a sample*

Accepted true value = 100.0%.

Analyst 1	Analyst 2
99.3%	99.1%
99.4%	100.2%
99.1%	99.8%
99.3%	99.5%
99.6%	100.4%
Mean 99.3%	99.8%
Standard deviation ±0.2	±0.5

The phase of the sample (gas, liquid, solid) or other characteristics such as corrosivity may exert an influence. The value of the sample, either intrinsic or due to collection costs, may determine whether destructive or non-destructive techniques are utilised. Obviously, non-destructive techniques enable several replicate analyses to be made, several analytes to be determined or the sample to be archived following analysis. The *reliability* of the procedure, instrument or technician, and the *simplicity* of the technique (i.e. can the procedure be automated or must a skilled technician be in constant attendance) must be assessed. These two criteria become extremely important for analyses to be performed during extended field operations (e.g. at sea for months during an oceanographic cruise) or for developing remote sensing methodologies.

Whether the analysis must be continuous or batch (discrete analyses sometime subsequent to sample collection) and the time required for the analysis must be considered simultaneously. Continuous monitoring of an analyte generally requires real time analyses. For instance, water treatment works determine the pH and residual chlorine levels as the water leaves the plant. Statutory regulations define water quality standards. Instantaneous analysis of the effluent allows immediate adjustments to chemical dosing to ensure that the water quality criteria are always met. Generally the effluent discharged from industrial sites into adjacent waters cannot be constantly monitored for constituents such as heavy metals. Thus, batch analyses are acceptable. Similarly, many environmental studies of natural waters would rely upon a field sampling programme followed by laboratory analyses at a later date.

Both capital and running costs must be considered in selecting a technique. Economic constraints may limit the options for several of the other criteria.

These first two considerations of the analyst's role involve the planning stages of the analysis. Thereafter, the final task is the analysis itself. This is briefly outlined in the following section.

4.1.2 Stages in an analysis

Any analysis will involve a number of steps, starting with the sample collection and following through to evaluating the data. All these stages are equally important, and the general principles are outlined below. Subsequent sections will consider each step in greater detail.

Sampling

This is a crucial but often ignored aspect of an analysis. A good sample must be obtained otherwise the ensuing analysis, possibly time-consuming and costly, becomes pointless. The sample presented for analysis must be free from contamination. In considering sea water, this presents no problem for the determination of Na^+ with a concentration of $11.8\,g\,dm^{-3}$. However, for Pb^{2+} with a concentration of only $10\,ng\,dm^{-3}$, exceptional care and expertise is necessary for a reliable result. Often the sampling procedure requires the sub-sampling of a large sample. This sub-sample must therefore be representative of the whole sample. This may be achieved for inhomogeneous material by creating a composite sample. A representative sample of tap water used domestically may best be obtained by ensuring that each time the faucet is turned on, a small aliquot is simultaneously withdrawn separately into a collection device. Inhomogeneous solids can be ground to a common grain size and the sample then split successively until an appropriate specimen size is obtained.

Preservation and pre-treatment

Analyses rarely can be performed immediately upon sample collection. Often storage is required, and the analyst must ensure that the integrity of the analyte is maintained. Accordingly, natural water samples for trace metal analytes are generally acidified to prevent adsorption onto the container walls. Some solutions may need to be chemically 'fixed' (i.e. with chloroform or mercuric chloride), or stored in cool (4°C) and dark conditions to prevent continued biological activity. Many samples must be protected from atmospheric oxidation.

The most common pre-treatment for liquid samples involves the separation of solid and liquid phases. This is achieved using either filtration or centrifugation. Pre-treatment for solid samples usually involves dehydration. This is done in order that the analyte can be expressed on a dry weight basis. Tissue and sediment contain variable amounts of water. This is removed by heating in a conventional oven or a microwave oven. Lyophilisation (freeze-drying) may also be utilised. Solid samples are often dissolved prior to analysis. This may be achieved with strong acids. Hydrofluoric acid, HF, is necessary for the dissolution of aluminosilicate material, while organic matter must be oxidised by HNO_3 or $HClO_4$. Alternatively, a strong base digestion such as tetramethyl ammonium hydroxide, $(CH_3)_4NOH$, may be suitable for solubilising organic matrices.

Separation and pre-concentration

Separation may be necessary to remove the analyte from potentially interfering substances. This is often used prior to analysis by atomic absorption spectrometry. It is, of course, absolutely essential to separate constituents to be analysed by a non-specific detector, as would be the case in most forms of gas–liquid chromatography.

Pre-concentration is carried out in order to increase the concentration of the analyte to a level at which it can be determined and often to separate it from interfering substances. This process may inherently involve a separation procedure such as solvent extraction or the use of chelating resins. Co-precipitation may be used in some instances. Precipitation of $Fe(OH)_3$ added to sea water quantitatively removes vanadium. Finally, an analyte may be pre-concentrated by simply evaporating off the solvent. However, any potential interferents are likewise pre-concentrated.

Measurement of the analyte

Several different methods are available for the measurement of the analytes. Some are listed in Table 4.2. Most of these techniques are outlined in Section 4.4.

Expressing the results

The result of an analysis comprises a numerical value and a notation of concentration units. The number given should be recorded with the appropriate number of significant figures. The number of significant figures can be designated as the number of digits necessary to express the result of a measurement in a manner consistent with the precision of the measurement. This will be the number of digits known with certainty and the first uncertain digit. For any given number, the last digit is taken to mean ± 1 unless otherwise specified.

The following numbers all have three significant figures:

$$0.417, \ 4.17, \ 417, \ 4.17 \times 10^5$$

Note that the exponential term does not affect the number of significant figures. Similarly, the zero in the first number is not significant. This is most obvious if the number is expressed in scientific notation as 4.17×10^{-1}.

The digit zero may be significant when falling between two other numbers or when it occupies the final digit after a decimal point. The following numbers *all* have three significant figures:

$$0.420, \ 0.0690, \ 81.0, \ 420., \ 7.30 \times 10^{-4}$$

Zeros used to define the location of the decimal point are themselves not significant. Hence, 420 with no decimal point has only two significant figures but 420. exhibits three significant figures. Expressing the values in scientific notation should clarify any ambiguity.

For expressing log term quantities such as pH, the number of decimal places expressed equals the number of significant figures in the original

Table 4.2. *Various techniques suitable for the quantification of an analyte*

Category	Technique
(1) Gravimetric	precipitation electrodeposition
(2) Titrimetric	precipitate formation acid–base compleximetric redox
(3) Energy absorption	molecular spectroscopy (UV–visible) atomic absorption spectroscopy infra-red spectroscopy
(4) Energy emission	atomic emission spectroscopy X-ray fluorescence electron spectroscopy (XPS, PES) atomic and molecular fluorescence
(5) Electroanalytical	potentiometry polarography voltammetry
(6) Chromatographic	thin layer chromatography gas–liquid chromatography high performance chromatography
(7) Others	mass spectrometry radiochemical techniques kinetic methods thermal analysis

non-exponential form. Therefore, the pH of the solution where $\{H^+\} = 1.93 \times 10^{-5}$ is 4.714 rather than 4.71. Conversely, the hydrogen ion activity for a solution of pH 11.23 is 5.9×10^{-12} and not 5.888×10^{-12}.

The number of significant figures that can be expressed following addition or subtraction will be limited by the number with the fewest decimal places given. This is shown in the following examples:

40.9	10.23	37.943
104.22	284.9	−15.63
1.81	1.479	———
82.4	713.65	22.31
———	———	
229.3	1013.3	

Multiplication and division requires more consideration in defining the appropriate number of significant figures. Generally the result can be expressed with the same number of significant figures as the *key number*, that number exhibiting either the fewest significant figures or the least precision. For example:

$$\frac{35.87 \times 903}{106.4 \times 0.6521} = 467$$

This result is understood to be 467 ± 1. The key number 903 contains only three significant figures. The precision of this number is therefore 1 in 903 and, on this basis, the precision of the final result is

$$\frac{1}{903} \times 467 = \pm 0.5$$

The result as expressed, namely 467, is therefore correct with respect to the overall precision. Consider the following very similar calculation:

$$\frac{35.87 \times 903.0}{106.4 \times 0.6521} = 466.8$$

As expressed, this implies 466.8 ± 0.1. The key figure in this calculation is 106.4, with a precision of 0.1 in 106.4 or 1 in 1064. Note that the precision of the other numbers is greater: 1 in 3587, 1 in 9030, 1 in 6521. The precision of the final result therefore would be:

$$\frac{1}{1064} \times 466.8 = \pm 0.4$$

Accordingly, the result for this calculation must be recorded as 466.8 ± 0.4.

Concentrations may be expressed in a range of units, e.g. amounts can be specified in $mol\,dm^{-3}$ or, as preferred by many thermodynamicists, $mol\,kg^{-1}$. But statutory limits are often stated in terms of weight per volume, i.e. the maximum admissible concentration of lead in drinking water is $50\,\mu g\,dm^{-3}$. Results for solids are best given as weight ratios on a dry weight basis. Gas concentrations may be given as the following ratios: weight to volume, weight to weight, or volume to volume. The atmospheric concentration of CO_2 on a volume:volume basis is about 340 ppm (see Chapter 1). This is calculated for a weight: volume ratio at STP (standard temperature, 0°C, and pressure, 101.3 kPa) to be:

CO_2 concentration is 340 ppm (v/v)
volume in $1\,m^3$ ($1000\,dm^3$) is therefore $3.40 \times 10^{-4}\,m^3$

$$n = \frac{PV}{RT} = \frac{101.3 \times 10^3 \times 3.40 \times 10^{-4}}{8.314 \times 273.15}$$

$$= 1.52 \times 10^{-2}\,mol$$

weight of $CO_2 = 1.52 \times 10^{-2} \times 44.0$

$$= 0.669\,g\ or\ 669\,mg$$

(Note that $R = 8.314\,J\,K^{-1}\,mol^{-1}$ and $1\,Pa = 1\,kg\,m^{-1}\,s^{-2} = 1\,J\,m^{-3}$; molecular weight of $CO_2 = 44.01$.)

Therefore, the concentration of CO_2 in the atmosphere at STP is $669\,mg\,m^{-3}$. This result may also be expressed on a weight:weight basis:

CO_2 concentration is $669\,mg\,m^{-3}$
density of dry air at 0°C is $1.293\,g\,dm^{-3}$
$1\,m^3$ of air weighs $1.293\,kg$

$$CO_2\ concentration = \frac{669 \times 10^{-3}\,g}{1.293 \times 10^3\,g}$$

$$= 517 \times 10^{-6}\,g\,g^{-1}$$

Thus, as a weight:weight ratio, the CO_2 concentration in the atmosphere is $517\,\mu g\,g^{-1}$.

The final role of the analyst is to evaluate the data. Having expressed a result in a form that is numerically correct and with acceptable notation of units, the analyst must decide whether the value obtained is reasonable. Thereupon the analyst may accept or reject the data. Data rejection should be kept to a minimum and based on objective reasoning rather than the analyst's whims. This is considered in the following section. Regardless of whether the data is accepted or rejected, the analyst may choose to repeat the analysis in order to improve the precision.

4.1.3 Errors and handling small data sets

Errors in an analytical procedure will affect the accuracy and precision of the measurement. *Systematic errors*, also termed determinate errors, are errors which are determinable. This means that they can be discovered and corrected. Systematic errors can arise due to faults in the equipment or analytical procedure. Poor calibration of an instrument or choosing the wrong colour as the end point for a titration while using a visual indicator are examples of such errors. Systematic errors will bias the results and so influence the accuracy of the analysis.

Random errors, also known as indeterminate errors, are unavoidable. They cannot be predicted and accordingly cannot be corrected. Such errors are associated with the uncertainty in every physical measurement. Examples are reading the scale on a burette to ± 0.01 or ± 0.02 cm^3 and colour variations at the visual end point of a titration. Random errors are accidental in nature and so several measurements will define a Gaussian curve. This normal distribution about a mean value is characterised by the standard deviation from the mean. Hence, random errors influence the precision of an analysis.

The *absolute error* is the difference between a measured value and the accepted true value. This difference may be termed the *mean error* if the measured value is the average of several measurements. The *relative error* is the absolute or mean error expressed as a percentage of the true value. For example:

$$
\begin{aligned}
\text{true weight} &= 1.73 \text{ g} \\
\text{mean measured value} &= 1.62 \text{ g} \\
\text{mean error} &= -0.11 \text{ g} \\
\text{relative error} &= \frac{-0.11}{1.73} \times 100 = -6.4\%
\end{aligned}
$$

As indicated previously, several replicate measurements of a quantity will define a Gaussian curve. This may require more measurements than is generally feasible or sometimes possible. Environmental chemists often deal with a sub-set of data that is meant to be representative of the whole population. Accordingly, handling small sets of data requires special consideration as commonly utilised statistical tests become meaningless when only few data are available.

Table 4.3 presents data, arranged in rank order, for the analysis of iron in a coastal marine sediment. These data are utilised to illustrate difficulties that may be encountered when only a limited number of measurements are available. The *mean* is the average value for all the measurements. This is calculated by summing all values and dividing by n, the number of measurements made. Alternatively, the *median* may be evaluated; this is the middle result for an odd number of measurements or the average of the middle pair of results for an even number of measurements. For the data in Table 4.3, the mean and the median are 3.15% and 3.12%, respectively. It should be noted that for small numbers of measurements, the median may represent the true result better than the mean does. This is because the median is less influenced by an outlying value. The *range* is the difference between the maximum and minimum measurements. For small data sets, the *standard deviation*, σ, (defined in Table 4.3) can be calculated from the range using a constant, k, (known as the deviation factor), as given in Table 4.4.

$$\sigma = k \times \text{range}$$

Thus, for our data, where $n = 7$ and range $= 0.30$:

$$\sigma = 0.37 \times 0.30$$

$$\sigma = 0.11$$

An extraneous value will affect the mean, as indicated above, but also leads to a high range and a relatively high standard deviation. This means that the precision is adversely influenced. Therefore, the analyst may wish to reject an outlying value. There should be some justification for this, otherwise the analyst runs the risk of rejecting good data and thereby biassing his results. Of course, it should be stressed that data must be rejected where a known error has occurred. Examples of this might be spilling a solution during a quantitative transfer or overshooting an end point during a titration. Otherwise outlying values may be rejected by using the *Q-test*. The *Q*-test examines the likelihood of a value from one end or the other end of the normal distribution (i.e. actually part of the whole population of measurements) to occur by chance in only a small sub-set of the whole population. Such a chance occurrence may distort the central tendency of the data.

For the Q-test, measurements $(X_1, X_2 \ldots X_n)$ are arranged in rank order. The ratio, Q, is defined to be:

$$Q = \frac{(X_2 - X_1)}{(X_n - X_1)} \quad \text{or} \quad Q = \frac{(X_n - X_{n-1})}{(X_n - X_1)}$$

Table 4.3. *A small data set for the analysis of iron in a coastal marine sediment*

Results	3.36%
	3.20%
	3.15%
	3.12%
	3.10%
	3.09%
	3.06%

Statistical parameters	Before Q test	After Q test
Number of measurements	7	6
Mean	3.15	3.12
Median	3.12	3.11
Range	0.30	0.14
Standard deviation*	0.10	0.06
Percentage of iron	3.12 ± 0.10	3.11 ± 0.06

* Standard deviation is a measure of the spread of the data about the mean. It is defined by the following expression:

$$\text{standard deviation, } \sigma = \left[\frac{\sum\limits_{i=0}^{\infty} n_i(\bar{d} - d_i)^2}{\sum\limits_{i=0}^{\infty} n_i - 1} \right]^{1/2}$$

where n_i is the number of measurements, d_i, and $\bar{d}$ is the mean value given by

$$\text{mean, } \bar{d} = \frac{\sum\limits_{i=0}^{\infty} n_i d_i}{\sum\limits_{i=0}^{\infty} n_i}$$

This depends on whether the outlying value is the highest (X_n) or lowest (X_1) measurement. If the value of Q calculated here exceeds the rejection quotient Q_{90} for n measurements, as tabulated in Table 4.4, then the outlying value can be rejected with a 90% confidence. This means that, while the outlying value is a part of the whole population of measurements, there is only a 10% chance that it will occur in that small number of measurements. The implication here is that the likelihood of two outlying values occurring by chance in a small data set is negligible. Thus, the Q-test can only be applied to reject data once to any set of measurements.

For our results in Table 4.3, the value 3.36% is tested for rejection:

$$Q = \frac{3.36 - 3.20}{3.36 - 3.06}$$

$$= \frac{0.16}{0.30}$$

$$= 0.53$$

Note that Q_{90} for seven measurements is 0.51 (from Table 4.4). Thus, this extraneous value of 3.36% can be rejected. The overall effect on the statistical parameters is shown in Table 4.3. The mean and median now correspond closely. The range is greatly reduced and the precision much improved.

Table 4.4. *The deviation factor, k, and rejection quotient, Q_{90}, for small numbers of measurements*

Number of measurements, n	Deviation factor, k	Rejection quotient, Q_{90}
2	0.89	—
3	0.59	0.94
4	0.49	0.76
5	0.43	0.64
6	0.40	0.56
7	0.37	0.51
8	0.35	0.47
9	0.34	0.44
10	0.33	0.41

From Dean & Dixon (1951).

4.1.4 Sensitivity and detection limit

Many analytes in environmental samples are present at only very low levels of concentration. This necessitates the use of very sensitive methods. The degree of response of the method to an analyte may be described in two ways:

Sensitivity

This is the concentration or mass of analyte required to give a specified level of response from the analytical method. For example, in spectroscopic absorption methods, the sensitivity is usually defined as the concentration of analyte (in $\mu g\,ml^{-1}$) required to give a 1% absorption of incident light. It thus provides a means of comparing responses to different elements or analyte species.

Detection limit

This is the smallest mass or concentration of analyte which can meaningfully be detected. This is usually influenced primarily by the *blank*, or the *baseline noise* of the method. The blank is the amount or concentration of analyte found by analysing an analyte-free sample of, for example, deionised distilled water in the case of water analysis. It is thus a measure of the contamination introduced from apparatus and reagents during the analytical procedure. It is not so much the absolute magnitude of the blank which determines the detection limit, as the variability of the blank. This can be exemplified by the two following hypothetical procedures of comparable sensitivity. *Method A* for the analysis of lead has a blank of 12.00 ± 0.01 (standard deviation) ng of lead, whilst *method B* has a blank of 1.0 ± 0.2 (standard deviation) ng of lead. Suppose a sample containing 0.1 ng of lead is analysed. In method A, it will give a response of 12.10 ng, which may readily be distinguished from 12.00 ng if both blank and sample have a standard deviation of 0.01 ng. In the case of method B, the sample response is 1.1 ± 0.2 ng, whilst the blank gives 1.0 ± 0.2 ng. These are not readily distinguished as the inherent variability in each measurement due to random error is so great. Thus method A has the better (lower) detection limit, despite having the higher blank. In fact, one definition of detection limit (there are several) is that

$$\text{detection limit} = 3\sigma \text{ (blank)}$$

where σ (blank) is the standard deviation of the blank. Thus methods A and B would have detection limits of 0.03 and 0.6 ng lead, respectively.

Some methods of analysis give no blank value as such, but measure a sample signal as a deviation in a fluctuating baseline (e.g. an atomic absorption spectrophotometer). In this case, the detection limit may be defined as three standard deviations of the baseline noise level (i.e. random fluctuations of the baseline).

In many cases the absolute standard deviation of the blank increases with the absolute magnitude of the blank. This is hardly surprising, and one consequence is that in most methods a low blank accompanies a low detection limit, as long as the method is of high sensitivity. Thus in environmental trace analysis, it is always advisable to take every reasonable precaution to avoid contamination as this enhances the attainable detection limit as well as avoiding collection of spurious data.

Whilst the detection limit seeks to define the smallest mass or concentration of analyte which can be detected with confidence, the *limit of quantitation* describes the lowest level at which that mass or concentration may be quantified with confidence. It is usually taken to be 10σ (blank).

4.2   Sample collection and storage

Collection of sound environmental chemical data starts outside the analytical laboratory. There is a great deal of skill in selecting an appropriate sample and collecting it without alteration, i.e. without contamination or depletion of the analyte of interest. Description of the selection of sampling sites is outside the scope of this text, but an introduction to the techniques of sample collection and subsequent storage prior to analysis is given below.

4.2.1 Air sampling

Air samples vary greatly according to the specific sampling application, but most contain the components shown in the general air sampling train (Figure 4.1).

The inlet system should be inert with respect to the analyte of interest, and commonly inlets are made of Teflon, glass or stainless steel. The siting of the air inlet itself in free air away from any solid surfaces is very important if a representative sample is to be taken. Most samples then include a particle filter. If the main aim is to collect a sample of airborne particles (known as aerosol), the nature of the filter material is very important. It should be a highly efficient particle collector, chemically inert and have a low background content of the analyte of interest. Commonly used filter materials which fulfil at least some of these criteria are glass fibre, and cellulose ester and Teflon membranes. Typical pore sizes for the membrane filters are around 0.5 μm, but, because of the complex mechanisms of particle collection, even the very smallest particles (down to a few nanometres in diameter) are collected with high efficiency.

If the analyte of interest is a gas, the particle filter will be followed by a gas collector or sensor. This may take the form of a wash bottle containing a liquid reagent, or may, for example, be an adsorbent porous polymer, dependent upon what is to be collected. To give a simple example, sulphur dioxide may be collected in a dilute hydrogen peroxide solution, where it forms H_2SO_4 which may be analysed by titration of acidity, measurement of sulphate or solution conductivity:

$$SO_2 + H_2O_2 \rightarrow H_2SO_4$$

The air is drawn through this system by a pump which is always used to suck, rather than blow, the air, as passage of air through the pump may alter its composition. Flow rates vary, but typically lie within the range 0.5–20 l min^{-1} for 'low volume' samples and are around 1–2 m^3 min^{-1} for 'high volume' samples.

In order to quantify the concentration of the trace analyte in the air, it is essential to know how much air has passed through the system. This may be achieved by use of a gas meter, which measures an integrated air

Figure 4.1. Schematic diagram of an air sampling train.

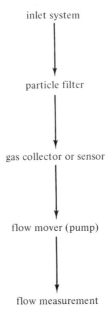

inlet system

particle filter

gas collector or sensor

flow mover (pump)

flow measurement

volume, or a flow rate meter, such as a rotameter, which gives a rate of flow, which may be converted to an air volume by multiplication by the duration of sampling.

Some commercially produced gas analysers have all of the above components built in, with a continuous sensor of some type used to measure the concentration of analyte. An example is the chemiluminescent analyser for nitrogen oxide, in which the air to be analysed is pumped continuously through a reaction chamber, in which it is mixed with ozone, produced within the instrument. Any nitrogen oxide in the air reacts with ozone to form a light-emitting excited state of nitrogen dioxide (denoted NO_2^*).

$$NO + O_3 \rightarrow NO_2^* + O_2$$

$$NO_2^* \rightarrow NO_2 + h\nu$$

Measurement of the intensity of emitted light by means of a photomultiplier tube gives a specific analysis of nitrogen oxide. Most instruments contain also a converter in which atmospheric nitrogen dioxide is broken down to nitrogen oxide, which is also analysed by the above reaction. A schematic diagram of the instrument appears in Figure 4.2.

Figure 4.2. Schematic diagram of a chemiluminescent analyser for oxides of nitrogen.

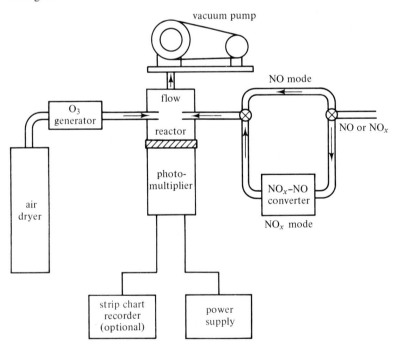

4.2.2 Water sampling

When sampling water, it is necessary to be aware that the composition of flowing water may change quite rapidly with time and that the composition of all but the most turbulent waters may vary with depth. Any sampling strategy will take account of these considerations. There is also a 'surface microlayer' which can be enriched in some trace species, especially those with surface active properties. It is thus necessary to fill the sampler at the known depth, and preferably to open, close and seal it at that depth. Commercially available devices will fulfil this function.

One of the biggest problems in water sampling and subsequent storage is the preservation of sample integrity prior to analysis. Problems arise for three main reasons.

(1) Leaching of contaminants from the vessel walls into solution. This cannot generally be prevented entirely, but may be minimised by selection of appropriate container materials.

(2) Precipitation and adsorption of trace components from the water. This is a problem, particularly with trace metals, and may in this case be overcome by acidification of the water prior to storage. This addition of acid will alter the sample, thus precluding some analyses, and may exacerbate the problem of leaching from the container walls outlined above.

(3) Microbiological activity may alter concentrations of essential nutrient species, especially nitrate and phosphate. This can usually be overcome by pre-filtration of the sample to remove microorganisms, or by storage at 4°C to minimise biological activity.

In this context, it should be mentioned that some waters, most specifically underground waters and deep ocean waters, are not in equilibrium with atmospheric gases. Exposure to the atmosphere will cause a change in the sample towards equilibrium, with consequent changes in pH, *Eh* (see Section 2.5.4) and carbonate levels. This type of sample may require analysing *in situ*, or before appreciable exposure to the atmosphere. The problems of redox potential changes are considered more fully in the context of aquatic sediments.

4.2.3 Aquatic sediments

The nature of the sedimentation process is crucially dependent upon the energy of the overlying waters. In a fast flowing energetic stream, fine-grained material will remain suspended, and only large gravel or pebbles will remain static on the stream bed. At the other extreme, a large deep lake has little horizontal flow of water, especially at depth, and virtually no turbulent or translational motion which will keep sediment grains in suspension. In this low energy environment, even fine-grained material will settle, forming a layer of bottom sediment. Such sediments may be little disturbed by chemical or biological agents and hence retain a stratified record of the depositional history of the lake basin. Figure 4.3 illustrates the variation in metal levels of sediments taken from lakes in North-West England and includes an estimate of sediment depositional age. Section 5.5 explains in greater

Figure 4.3. Vertical profiles of copper, lead and zinc in the sediments of Wastwater (WA), Killington Reservoir (KR) and Hollingworth Lake (HL) in north-west England. From Hamilton-Taylor (1983). * Depth scale constant for all samples; time scale varies due to different sedimentation rates.

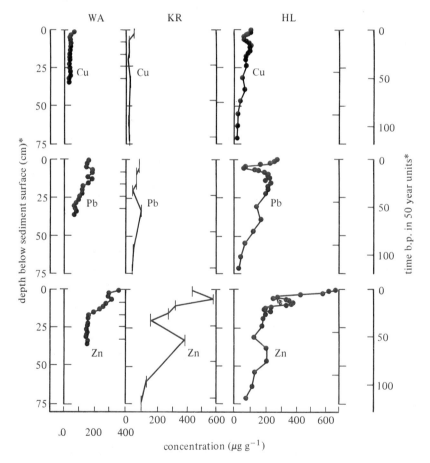

detail the factors influencing sedimentation processes and subsequent changes in sediment properties in relation to the marine environment.

Because sediments can retain a record of depositional history, it is often desirable to collect a depth profile sample, known as a *core*. There are many types of corers and borers designed to undertake the difficult task of collecting core samples from under water; a description of their function is beyond the scope of this text. Collection of inter-tidal sediments at low tide is an easier task as the sediment is exposed. Under such circumstances, a core tube, conveniently fabricated from a 30 cm or longer length of unplasticised-pvc drainpipe and with a sharpened cutting edge, is hammered into the sediment and then removed by digging out the surrounding sediment. Careful extrusion from the tube exposes an intact core.

Since aquatic sediments normally contain microbiota which consume available oxygen, it is normal for only the surface layer of an inter-tidal sediment to be *aerobic* (except in the case of large-grained, sandy sediments). Lower layers are entirely depleted in oxygen (*anaerobic* conditions). This may have a consequence for handling of the sediment if the chemical constituents of interest have an active redox chemistry. For example, iron exists predominantly as $Fe(III)$ in the aerobic section and as $Fe(II)$ in the anaerobic, or *anoxic*, part. As soon as the anaerobic section is exposed to the air, oxidation of $Fe(II)$ to $Fe(III)$ progresses fairly rapidly. Thus, if the redox chemistry is to be preserved, the sediment must be sealed within the core tube and subsequently extruded and handled in an inert nitrogen atmosphere in a glove box or bag. Some sediments are in contact with overlying anoxic waters and are liable to be anoxic through their entire depth.

Once the sediment has been collected, it is normal to divide it into depth fractions for subsequent physical and chemical analyses. Many sediment properties are grain-size dependent, and analysis by sieving through standard mesh sizes is commonly utilised to determine coarse grain sizes; other methods based upon light scattering, or sedimentation properties, are used to characterise fine-grained material.

Sediments are commonly stored within the core tube used for collection, with maintenance of a temperature of 4°C to minimise biological activity.

4.2.4 Soils

Undisturbed soils contain a number of layers, known as *horizons* of differing composition (see Section 5.4). Thus, like aquatic sediments, the chemical make-up is depth dependent, and sampling strategies must allow for this. The use of core tubes is possible in a soil relatively free of stones, but where such obstructions exist other depth sampling techniques must be used. These include the use of *augers*, which are devices screwed into the ground, and the technique of simply digging a suitable hole which exposes a clean depth profile from which samples may be removed.

Apart from depth inhomogeneities, soils show inhomogeneities of other types. Surface soil commonly varies considerably in composition over small horizontal distances, and frequently a number of samples must be taken from, say, a polluted site to gain an impression of the typical concentration of a pollutant element. A further complication arises from the fact that soils may contain large aggregate grains, and, to achieve a homogeneous sample of soil which may be successively sub-sampled with the same composition, it is necessary to grind a dried soil to a fine powder; commonly a grain size of below 150 μm is recommended. Materials of greater than 2 mm diameter (i.e. stones, twigs, etc.) are normally discarded entirely.

Soils may be air-dried simply by exposure to ambient air, or oven-dried at an elevated temperature. The latter will achieve a lower moisture content, but may also volatilise some organic material. There is no universally accepted procedure for drying soils. Once dried, a soil may be stored in a desiccator with little subsequent chemical change.

4.3   Pre-treatment techniques

4.3.1 Filtration of water samples

Almost all natural waters contain both dissolved and suspended chemical constituents. Before analysis of the sample, it is usual to separate these two phases and analyse them independently. Most commonly, filtration is the technique used, and by convention a 0.45 μm pore size filter is generally employed.

In recent years it has become recognised that to consider the constituents of a water as solely dissolved or suspended, with a clear division of the two, is a gross over-simplification, and that chemical components frequently cover a continuum of sizes from truly dissolved free and solvated ions, through high molecular weight organics and colloidal suspensions to particulate material, itself covering a wide range of sizes. In the case of metals, it is quite possible for a metal to be present within a water in all of these chemical associations to some degree, and this is illustrated in Table 4.5. Thus the 0.45 μm filtration does not give the clear-cut division of the sample envisaged originally, but is still very widely used as it is a convenient means of dividing the sample. The fractions generated are normally termed 'dissolved' (passes the 0.45 μm filter) and 'suspended' (collected on the filter), although it is accepted that such fractions are operationally defined (i.e. determined solely by the operation of 0.45 μm filtration, rather than by a clear-cut physico-chemical division).

4.3.2 Pre-concentration methods

Since many substances are present at only very low levels in environmental media, it is often necessary to pre-concentrate them in some manner prior to chemical analysis. This may be achieved in a wide variety of ways, dependent upon the type of sample and the nature of the analyte.

The latest techniques for analysis of metals are of sufficient sensitivity for direct assay of many metals in natural waters. Where older instrumentation is employed, or for elements present at a very low abundance, a pre-concentration may be required. Where little chance of

Table 4.5. *Metal species occurring in natural waters in relation to size association*

Typical size range (nm)	Metal species	Example	Phase state
<1	free metal ions	Pb^{2+}, Cu^{2+}	dissolved
1–10	inorganic ion pairs, inorganic complexes, low molecular weight organic complexes	$CdCl_4^{2-}$ Pb-fulvates	dissolved
10–100	high molecular weight organic complexes	Cu-humates	colloidal
100–1000	adsorbed on to inorganic colloids (or complexation by surface-adsorbed humics); associated with detritus	$Co-MnO_2$ Pb-FeOOH	colloidal
>1000	adsorbed on to living cells; associated with mineral solids and precipitates	Cd-clay $2PbCO_3 \cdot Pb(OH)_2$	particulate

sample contamination exists, the crudest form of pre-concentration, simple evaporation to a smaller volume by heating, may be effective and has commonly been employed with fresh waters. Alternatively, a metal-chelating reagent (e.g. ammonium pyrolidine dithiocarbamate – APDC) may be added and the complexed metal extracted into a small volume of organic solvent (e.g. methyl isobutyl ketone – MIBK or a freon, $CFCl_2$—CF_2Cl). The metals are thus pre-concentrated by an amount equal to the ratio of volume of water sample and organic solvent, and indeed the use of organic solvents with flame atomic absorption spectrometry (see Section 4.4.5) gives a further enhance-ment in sensitivity relative to analysis of aqueous samples. In other cases the metal is re-extracted from the organic solvent into an acid medium which is then analysed.

Organic compounds in both air and water may be pre-concentrated by passage of the sample through a porous organic polymer or resin where the analyte is absorbed. Thus hydrocarbons in street air are commonly collected upon a porous polymer such as Tenax, from which they may be displaced by heating for a subsequent chemical analysis. In the case of water samples, organics collected on a polymer or resin are eluted by an appropriate solvent, rather than by heat. In an alternative technique, volatile organics may be purged from water by sparging with air (i.e. upward bubbling of air through the water sample). They are collected from the air onto a porous polymer, charcoal trap or cold trap, from which they may be solvent extracted or analysed in a manner analogous to that described for air samples (known as purge-and-trap). In this case the efficiency of purging from the water must be taken into account in the calculation of concentration.

4.3.3 Ashing and digestion techniques

Many environmental samples are not in a satisfactory form for direct introduction into an analytical instrument. Frequently, an aque-ous solution is required and thus some form of sample treatment which both dissolves the analyte and destroys associated matrix materials is highly desirable.

Dry ashing involves destruction of organic material by heating the sample, which may be, for example, of biological tissue, at around 500°C. This leaves the involatile constituents in an accessible and relatively easily dissolved state. This procedure, whilst useful for highly involatile analytes, can cause partial volatilisation of many elements and is rather out of favour. The alternative of low temperature ashing in an oxygen plasma is more effective, giving better retention of potentially volatile elements, but requires expensive apparatus.

The most commonly used procedures with tissue, soil, sediment and air filter samples involve digestion, or 'wet ashing procedures' for extraction of non-degradable analytes such as metals. The basic requirement is for a strong oxidising acid which will decompose organic material, whilst solubilising the analyte element. The most commonly used acid is nitric acid, HNO_3, which may be added to the sample in a beaker and evaporated to near dryness on a hotplate or sand bath. Two such treatments are normally sufficient to digest small samples of vegetation. The residue is then dissolved in dilute acid (say 0.1% to 5%) for subsequent analysis.

When stronger oxidising powers are required to destroy organic materials, it is common to use perchloric acid, $HClO_4$, in mixtures with nitric acid. This will give efficient destruction of for example, fish tissue and oils, which are not readily digested by nitric acid. Concerns over the safety of perchloric acid, which if used improperly can pose an appreciable explosion and fire hazard, limit the use of this reagent.

Soils and sediments generally contain some silicate materials (e.g. clays). The silicate lattices of these compounds are not broken down by oxidising acids, and to give a full breakdown, and thus to release fully the associated trace elements, it is necessary to digest such samples in a mixture containing hydrofluoric acid, HF. This breaks the silicate lattices and releases the silicon as volatile SiF_4. With heavily polluted samples it is rarely necessary to use hydrofluoric acid in the digestion as pollutant material does not enter the silicate lattice, and may be extracted with other acids such as nitric. On the other hand, in unpolluted samples an appreciable proportion of the trace metal burden may be within the silicate lattices, and HF digestion is essential if it is to be released. It is interesting to note that laboratory glass is a silicate and is thus subject to attack by HF. Digestions with this acid are carried out in inert Teflon (polytetrafluoroethene) beakers, and stringent precautions are taken to avoid contact of the acid with exposed human skin.

4.4   Analytical methods

4.4.1 Titrations

Although titrations are steadily being supplanted by more rapid and sensitive analytical methods, there are still many analyses of environmental samples which are commonly performed by titration.

The simplest type of titration in conceptual terms is the acid–base titration. This involves metered addition of an acid (or base) of known concentration to a known volume of a base (or acid) of unknown concentration (see Section 2.4.4 for a discussion of acid–base reactions). An indicator, which is a substance which changes colour at a known pH value (see Section 2.4.4 for a definition of pH) is used to determine when the added acid (or base) has neutralised the base (or acid) in the unknown sample. For example, in a titration of hydrochloric acid with sodium hydroxide, the indicator, litmus for example, changes colour at the point where exactly equivalent amounts of HCl and NaOH are present and reaction (4.1) has just gone to completion:

$$NaOH + HCl \rightarrow NaCl + H_2O \qquad (4.1)$$

This occurs at the neutral pH of 7, and addition of very small amounts of either HCl or NaOH to the mixture at pH 7 will cause very marked shifts in pH and a change in indicator colour.

All is not so simple, however. If, for example, the weak base calcium carbonate is titrated with hydrochloric acid, the reaction proceeds as follows:

$$CaCO_3 + 2HCl \rightarrow CaCl_2 + H_2O + CO_2 \qquad (4.2)$$

Firstly we note that, in contrast to reaction (4.1) which involves a 1:1 ratio of acid:base, reaction (4.2) involves a 2:1 ratio, which must be accounted for in calculating the concentration of our unknown solution. Secondly, if, using solutions of known concentration, exactly equivalent amounts of $CaCO_3$ and HCl are mixed, the resultant pH is *not* 7.0. This is because a $CaCl_2$ solution is slightly acidic due to hydrolysis, to form $Ca(OH)_2$ – a weak base (poorly dissociated) and HCl – a strong acid (highly dissociated), leading to an excess of H^+ over OH^- ions in the solution. To determine the equivalence point of the titration, it is therefore necessary to select an indicator which changes colour at an acidic pH corresponding to a dilute solution of $CaCl_2$.

The carbonate equilibria which exist in natural waters have been introduced in Section 2.4.4. It is often valuable to assess the capacity of a natural water to neutralise hydrogen ions, termed the *alkalinity* of the water, and this is done by titration of the water with a standard acid solution (by convention, HCl). The alkalinity of the water is due predominantly to the presence of natural dissolved Group I and II metal carbonates and bicarbonates, and at high pH to hydroxyl ions. These neutralise acid according to the following reactions:

$$OH^- + H^+ \rightarrow H_2O \qquad (4.3)$$

$$CO_3^{2-} + H^+ \rightarrow HCO_3^- \qquad (4.4)$$

$$HCO_3^- + H^+ \rightarrow H_2CO_3 \qquad (4.5)$$

By convention, the titration of alkalinity is carried out using two indicators. The first to be added is phenolphthalein, which changes colour at pH 8.3. Most natural waters are below this pH and thus have a zero *phenolphthalein alkalinity*. Some, however, are of higher pH, and the amount of acid added to the phenolphthalein endpoint is used to calculate the phenolphthalein alkalinity and corresponds approximately to the neutralisation of all hydroxyl and half of the carbonate within the sample.

A second indicator, methyl orange, is then added. This changes colour at pH 4.5, at which pH all alkaline species have been neutralised, and it thus permits calculations of the *total alkalinity* of the sample. The existence of two equivalence points is exemplified in Figure 4.4, which shows the change in pH when a standard solution of Na_2CO_3 is titrated with HCl. The calculation of alkalinity could be made complicated by the fact that at least three separate chemical species may have been titrated by the acid. In fact, for simplicity, it is imagined that all alkalinity arises from the presence of calcium carbonate, $CaCO_3$, and the calculations assume that the titration can be represented by reaction

Figure 4.4. Variation in pH when 25 ml of 0.10 M Na_2CO_3 is titrated with 0.10 M HCl. The sharp drops in pH correspond to the 'equivalence points'.

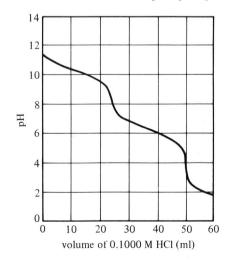

(4.2). Alkalinities are then expressed in units of $mg\,l^{-1}$ $CaCO_3$, or alternatively in $meq\,l^{-1}$ of added hydrogen ions.

Worked example
Titration of 100 ml of river water with 0.100 M HCl solution consumes 2.85 ml HCl to the methyl orange equivalence point. Calculate the total alkalinity of the sample.

The assumed chemical process is described by

$$CaCO_3 + 2HCl \rightarrow CaCl_2 + H_2O + CO_2$$

1 mol HCl neutralises 0.5 mol, or 50.05 g $CaCO_3$.

$$1\,l \text{ of } 1\,M \text{ HCl} \equiv 50.05 \text{ g } CaCO_3$$

Thus

$$2.85 \text{ ml of } 0.100 \text{ M HCl} \equiv 50.05 \times 2.85/1000 \times 0.1 \text{ g}$$
$$\equiv 14.3 \text{ mg } CaCO_3$$

This is in 100 ml river water, therefore

$$\text{total alkalinity} = 14.3 \times 1000/100$$
$$= 143 \text{ mg } l^{-1} \text{ } CaCO_3$$

Alternatively, 2.85 ml of 0.100 M HCl contain

$$2.85/1000 \times 0.1 \text{ g eq } H^+$$

as a 1 M solution contains $1\,mol\,l^{-1}\,H^+$. Thus

$$\text{alkalinity} = 2.85/1000 \times 0.1 \times 1000/100 \text{ eq } l^{-1}$$
$$= 2.85 \times 0.1 \times 1/100 \times 1000$$
$$= 2.85 \text{ meq } l^{-1}$$

Another type of titration still quite commonly used in environmental chemistry is the redox reaction titration. This is used in water chemistry, for example in the measurement of dissolved oxygen by Winkler titration or of chemical oxygen demand (COD) by dichromate oxidation. The basis of redox chemistry is introduced in Section 2.5.

The use of a redox titration can be illustrated by the measurement of chemical oxygen demand. In this method, a water or effluent sample is heated under reflux with acidic potassium dichromate. The dichromate provides an oxidising capacity which destroys organic matter, and, from the amount of dichromate reacted, it is possible to calculate the oxygen demand of the water or effluent. To determine the amount of dichromate remaining after the digestion under reflux, it is titrated with a reducing reagent, generally ferrous ammonium sulphate solution in the presence of an indicator which changes colour when all dichromate has been reacted in the titration (tris(1,10-phenanthroline)iron(II) sulphate).

The overall reaction of dichromate ($Cr_2O_7^{2-}$) and ferrous iron (Fe^{2+}) is as follows:

$$6Fe^{2+} + Cr_2O_7^{2-} + 14H^+ \rightarrow 6Fe^{3+} + 2Cr^{3+} + 7H_2O \qquad (4.6)$$

This overall reaction covers two redox processes:

$$6Fe^{2+} \rightarrow 6Fe^{3+} + 6e^- \qquad (4.7)$$

$$Cr_2O_7^{2-} + 14H^+ + 6e^- \rightarrow 2Cr^{3+} + 7H_2O \qquad (4.8)$$

Knowledge of the reaction stoichiometry and the volume and strength of ferrous solution allows calculation of the residual dichromate in the solution.

4.4.2 Electrochemical methods

Ion-selective electrodes are electrochemical devices which, when inserted in an aqueous solution, give a response which is specific or selective towards one particular ion in the solution. These are available for the determination of several ions, but by far the most familiar is the glass electrode used for measurement of pH.

The glass electrode consists of a silver/silver chloride electrode in dilute hydrochloric acid which is enclosed in an envelope of electrically conductive glass. The glass electrode is immersed in the analyte solution, together with a reference electrode. When the concentration of hydrogen ions in the analyte solution differs from that within the glass electrode, a potential difference exists between the two solutions. This difference is quantified by an electrometer connected to the silver/silver chloride (measures potential difference across the glass electrode) and reference (measures potential of external analyte solution) electrodes. According to the Nernst equation (Section 2.5.5), the potential difference is related to the activity of hydrogen ions in contact with the glass bulb by

$$E_g = \text{constant} + 2.3\,(RT/F)\log\{A_{H^+}\} \tag{4.9}$$

or at 25°C

$$E_g = \text{constant} - 0.059\,\text{pH} \tag{4.10}$$

Thus the potential is a linear function of pH, and the electrode sensitivity at 25°C is 59 mV (pH unit)$^{-1}$. Sensitivity is a function of temperature, and most pH instruments incorporate automatic temperature compensation. Before use, pH meters must be calibrated against buffer solutions of known pH.

Another widespread use of electrochemistry in chemical analysis is in *polarography* and related techniques such as *anodic stripping voltammetry*. The basic polarograph consists of a glass cell containing the analyte solution and two electrodes. One electrode consists of a pool of mercury in the bottom of the cell, and the other of a stream of small mercury drops (typically 2 to 6 s per drop) which are formed at the tip of glass capillary fed with mercury by gravity. The latter electrode, known as a dropping mercury electrode, has a continuously renewing surface due to the continual expansion and replacement of the drops. The solution in the cell contains the analyte (often a metal) in solution in a 'supporting electrolyte', which is a conducting ionic solution. A steadily increasing or decreasing voltage is applied between the electrodes and the current passing through the cell is monitored. The resultant polarogram (Figure 4.5) shows a slow increase in current with applied

Figure 4.5. A typical polarogram.

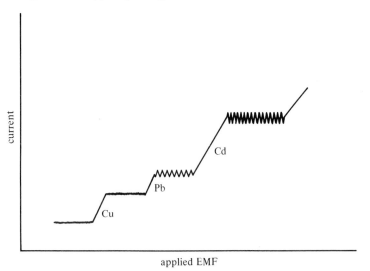

applied EMF

potential, upon which are superimposed far more rapid increases due to reduction of metal ions at specific potentials, which are diagnostic of the analyte. For example:

$$Cu^{2+} + 2e^- \rightarrow Cu^0 \tag{4.11}$$

The reduced metal is incorporated in the mercury as an amalgam. The magnitude of the steps are related to the concentrations of analyte ions, which are determined by the use of calibration solutions.

The limited sensitivity of polarography (about $1\ mg\ l^{-1}$ detection limit for most metals) has led to the introduction of the much more sensitive technique anodic stripping voltammetry (ASV), which allows many more elements to be determined, e.g. aluminium, titanium, vanadium, molybdenum. Once again, in ASV electrodes are used in a glass cell containing analyte in a supporting electrolyte (Figure 4.6). In this case, one electrode comprises a static mercury drop hanging from the end of a glass capillary from which it has been extruded. This is known as a hanging mercury drop electrode; a reference electrode completes the circuit. A constant negative potential is applied to the hanging mercury drop electrode for a determined time interval, during

Figure 4.6. Anodic stripping voltammetry cell using a hanging mercury drop electrode.

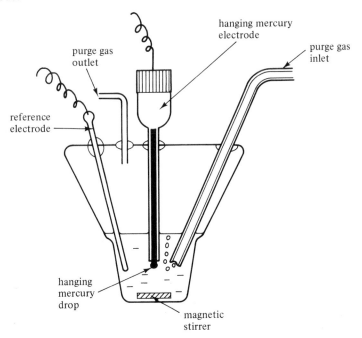

which electrodeposition of metal ions from the stirred solution occurs. Subsequently, a steadily increasing positive potential is applied to this electrode which causes sequential oxidation (or anodic stripping) of analyte metals back into the solution as their individual redox potential is reached. For example, for lead

$$Pb^{2+} + 2e^- \rightarrow Pb^0 \qquad \text{electrodeposition} \qquad (4.12)$$

$$Pb^0 \rightarrow Pb^{2+} + 2e^- \qquad \text{anodic stripping} \qquad (4.13)$$

Each analyte metal is stripped at a characteristic potential, the stripping current relating to the concentration of the ion in the initial solution. The accumulation of metal during the electrodeposition step leads to a great enhancement in sensitivity relative to simple polarography.

4.4.3 X-ray analytical methods

Two X-ray techniques will be described. These depend upon very different properties of X-rays and yield very different types of chemical information about a sample. Unlike most of the analytical methods described in this chapter, which require liquid or gaseous samples, both X-ray procedures are most readily applied to solid samples.

X-ray fluorescence (XRF)

When a metal is bombarded with accelerated electrons it will emit X-rays. The bombarding electrons knock atomic electrons from their orbitals, and when other electrons 'jump' from outer orbitals to replace those lost from inner orbitals (see Chapter 1) they lose energy by emission of radiation in the X-ray region. Since these effects arise from inner orbitals, valency electrons in outer orbitals have little influence on the energetics of the process, and the X-ray emission is characteristic of the chemical element rather than of any particular chemical form or compound of that element. If, instead of electrons, X-rays are used to bombard the metal, some absorption of the primary X-ray beam occurs, with an associated emission of fluorescent, secondary X-rays which are of lesser energy than the primary. The fluorescent X-rays occur at discrete energies which correspond to specific energy level transitions within the atom from which they arise. The energies, and thus wavelengths, of the secondary X-rays are characteristic of the emitting element and hence provide a fingerprint of the elemental composition of a sample. For practical reasons, measurements are generally limited to elements of atomic number 11 (sodium) or greater, but above this point almost all elements may be analysed.

By use of standards of comparable form and chemical matrix to the sample, it is possible to calibrate secondary X-ray intensities against the abundance of a particular element in a sample. X-ray fluorescence spectrometers are of two main types – wavelength dispersive, in which a crystal is used in a manner analogous to a prism to disperse the secondary X-rays according to their wavelength, and energy dispersive, in which the X-rays are characterised by a detector according to their energy without prior separation of wavelengths. The former technique is more sensitive, but the latter offers simultaneous analysis of many elements, and thus greater speed of analysis.

Many types of environmental samples may be analysed by X-ray fluorescence. These include rocks, soils and sediments in a dried and powdered state, suspended sediments from natural waters collected on a filter surface, and airborne particles collected on a filter. One great advantage is that the method is essentially non-destructive and hence the sample may be used for other analytical purposes subsequent to XRF examination.

X-ray powder diffraction (XRD)

Whilst XRF is used to analyse the elemental composition of a sample, XRD can identify specific crystalline chemical *compounds* with a high degree of certainty. It is based upon the fact that the atoms within a crystal are within a regular arrangement of planes which act as a 'diffraction grating' for X-rays. This is illustrated in Figure 4.7, which shows how X-rays may be 'reflected' from planes of atoms when the conditions of the Bragg equation (4.14) are met

Figure 4.7. X-ray diffraction by planes of atoms within a crystal.

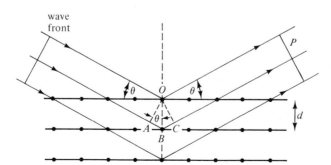

$$n\lambda = 2d \sin \theta \qquad\qquad (4.14)$$

in which

> $n = 1, 2, 3 \ldots$ is the order of reflection,
> λ = X-ray wavelength,
> d = inter-planar spacing,
> θ = angle of incidence.

The actual arrangement of a simple powder diffraction apparatus is shown in Figure 4.8. The sample takes the form of a mass of finely divided crystals in a random orientation which is rotated to maximise the number of orientations. At some orientation of the crystal, the conditions of the Bragg equation will be met and X-ray diffraction will occur, deflecting X-rays onto the film, or in more modern instruments a rotating gas-filled detector, at an angle of 2θ to the incident beam. Since a fixed wavelength, λ, is used, the angle 2θ is directly related to the inter-planar spacing d by equation (4.14). Each crystalline compound has its own set of 'd-spacings' and thus gives a characteristic diffraction pattern from which it may be identified using files of previously recorded diffraction data in conjunction with search manuals. Since the lines recorded by the X-ray film are narrow, it is possible to identify many individual components of a multi-component mixture by this technique. It must, however, be emphasised that the method is applicable only to materials with a well-defined crystal structure, and hence poorly crystalline or amorphous solids cannot be identified as they do not give clear diffraction patterns.

X-ray powder diffraction is commonly used to identify minerals in aquatic sediments. It is particularly well suited to use with clay minerals, many of which are of rather similar chemical composition, but have discrete crystal structures which provide characteristic diffraction patterns. The method has also yielded very useful data on metal com-

Figure 4.8. Simple apparatus for X-ray powder diffraction measurements.

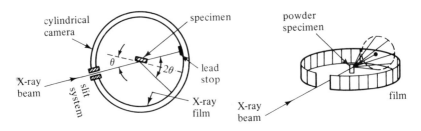

pounds in polluted air as well as on particulate salts containing sulphur and nitrogen.

4.4.4 Mass spectrometry

In a typical mass spectrometer (Figure 4.9) the analyte is introduced, usually in gaseous form, into the source, where it is bombarded with a stream of electrons, usually of energy 70 eV. (1 eV, or electron-volt, is the energy acquired by an electron when accelerated through a potential difference of one volt: $1\ eV \equiv 96.5\ kJ\ mol^{-1}$.) This causes the formation of positive ions by knocking electrons from the analyte atoms or molecules. These ions are accelerated out of the source by a high negative potential from whence they travel in a high vacuum through a magnetic field which curves the path of their motion. Their kinetic energy, $\frac{1}{2}mv^2$ (m = ion mass, v = velocity) is equal to the accelerating energy UQ, where U is the accelerating voltage and Q is the

Figure 4.9. Schematic diagram of a magnetic-focusing mass spectrometer.

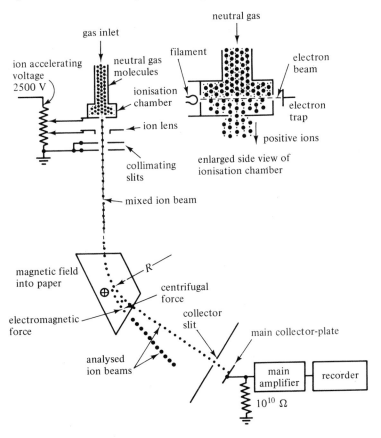

charge on the ion. Under normal conditions most ions acquire only a single positive charge, equal to the charge on an electron, e. Then,

$$\tfrac{1}{2}mv^2 = Ue \qquad (4.15)$$

If the ion is curved through a radius R, equating the force exerted by the magnetic field to the centrifugal force on the ion,

$$Bev = \frac{mv^2}{R} \qquad (4.16)$$

where B is the magnetic flux density. Then from equation (4.15)

$$\frac{m}{e} = \frac{2U}{v^2} \qquad (4.17)$$

Substituting for v from equation (4.16) gives

$$\frac{m}{e} = \frac{B^2R^2}{2U} \qquad (4.18)$$

Thus for any given value of magnetic flux density, B, the accelerating voltage, U, can be adjusted by bringing an ion of any mass to charge ratio, m/e, through radius R to the detector. In practice, the accelerating voltage can be ramped automatically to bring each ion of progressively higher mass into focus on the detector, thus generating a mass spectrum. An example appears in Figure 4.10.

If a monatomic element is introduced into the mass spectrometer, it would be anticipated that only one mass peak corresponding to the atom of that element would be observed. Commonly, however, elements are a mixture of different isotopes (see Section 1.1) and peaks are observed corresponding to each isotope, with intensities directly proportional to their relative abundances.

In environmental analysis, one common application of the mass spectrometer is to identify and quantify organic compounds. A typical application might involve analysis of organic pollutants in water. These are first extracted into a solvent and then, since many individual compounds are present, injected into a gas chromatograph to give a separation. The column effluent from the gas chromatograph is, after separation of the carrier gas, passed into the source of the mass spectrometer. A mass spectrum may then be generated corresponding to each peak in the chromatogram. A single organic compound will give a complex mass spectrum. The ion formed initially by removal of one electron from the organic compound, termed the molecular ion, is

commonly inherently unstable and decays to give other charged fragments. These fragments contribute to the mass spectrum and give a
pattern which is characteristic of, but rarely unique to, the particular
compound. This can be helpful, since, for example, all isomers (see
Chapter 3) of C_5H_{12} will give the same molecular ion, $C_5H_{12}^+$, but
differences will exist between the fragmentation patterns, and hence the
observed mass spectra, of the many different isomers. This may allow
identification. In more extreme cases in which, say, an alkanal and an
alkanol have the same molecular formula, separate identification of the
basis of the mass spectrum may be rather easy. Reference spectra are
retained upon a computer-based memory, and comparison of experimental and reference spectra is carried out automatically to provide a
tentative identification of an unknown compound. Figure 4.10 includes
the mass spectra of three compounds, all of molecular weight 126, and
two with an identical molecular formula, C_9H_{18}. It is evident that

Figure 4.10. Mass spectra of three compounds of identical molecular weight.
(M^+ = molecular ion.).

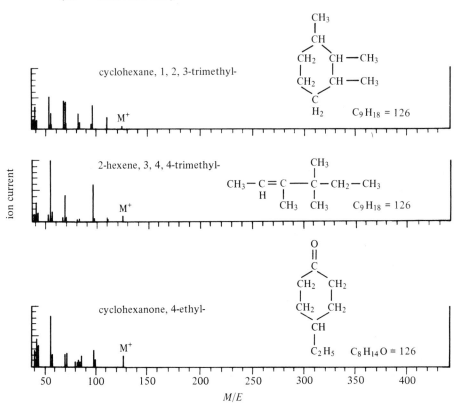

discernible differences exist in the three mass spectra. If a very high temperature plasma is used as the ion source, the mass spectrum comprises ionised atoms rather than molecular fragments.

Mass spectrometers are also widely used for determination of isotopic ratios of elements which may provide valuable insights into sources and environmental pathways of elements.

4.4.5 Absorption spectroscopy

The general principles and instrumentation are similar for molecular spectroscopy in both the UV–visible and infra-red wavelength regions, and for atomic spectroscopy. Figure 4.11 shows the basic layout of a simple single beam spectrometer. The instrument is first zeroed with a blank solution in the sample cell. The sample, which absorbs light at the analytical wavelength selected by the monochromator, is then introduced into the sample cell and the intensity of transmitted light is then measured by the detector. In double beam instruments, optical techniques are used to split the light beam from a single source into reference and sample beams which reduces the need for frequent zeroing as this may be carried out automatically using the reference light beam.

If the intensity of light incident upon the sample is I_0, and that transmitted through the sample is I,

$$\% \text{ transmission} = I/I_0 \times 100 \qquad (4.19)$$

and

$$\text{absorbance, } A = \log_{10}(I_0/I) \qquad (4.20)$$

Instrumental readout is normally in terms of one of these parameters. Calibration is with standard solutions of the analyte. Commonly it is found that absorbance is directly proportional to concentration, c, a relationship known as Beer's law. In fact, at low concentrations,

Figure 4.11. Schematic diagram of a single beam UV–visible spectrometer.

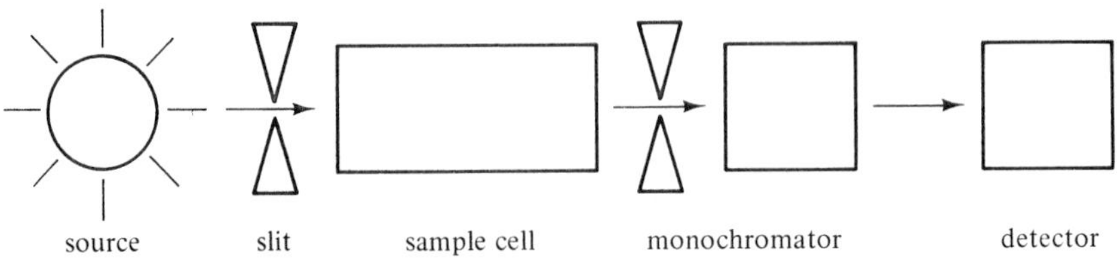

source   slit   sample cell   monochromator   detector

$$A = \varepsilon cl \tag{4.21}$$

in which l is the path length of the sample cell (the measured absorbance is directly proportional to l) and ε, the constant of proportionality, is known as the absorptivity or extinction coefficient, a characteristic of the light-absorbing substances being analysed. The percentage transmission is never linearly related to analyte concentration.

UV–visible absorption spectroscopy of molecules

Light emission or absorption by atoms and molecules in the UV–visible region ($\sim$180–700 nm wavelength) is due to transitions of electrons between energy levels (see Sections 1.1 and 1.2) within the atom or molecule. Molecular UV–visible spectra may be obtained in the gas or liquid phase; most practical measurements in environmental chemistry are made upon liquid phase samples. Under these conditions, molecular absorption spectra take the form of *band spectra* (as opposed to line spectra), which means that absorption of light occurs continuously over a range of wavelengths, rather than at single discrete wavelengths. This is illustrated by Figure 4.12. Analytical measurements are generally, but not exclusively, made at the wavelength of maximum absorbance, λ_{max}, so as to maximise sensitivity.

In environmental chemistry, UV–visible absorption spectroscopy is generally used to quantify the concentration of a substance known to be present. If, as usually happens, that substance does not have convenient light absorbing properties within the UV–visible wavelength range, it is necessary to use some chemical property of the substance to make a light-absorbing derivative, preferably in such a way that no other, potentially interfering light absorbing derivatives of other substances

Figure 4.12. UV–visible absorption spectrum of benzene in ethanol solution.

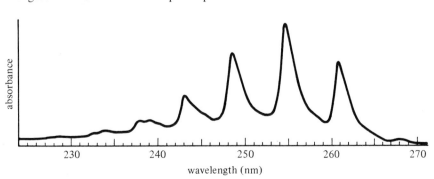

are formed. Hence the derivatisation reaction should be *specific* to the analyte of interest.

A simple example of environmental analysis using UV–visible spectrophotometry is the assay of ozone in air by bubbling a known volume of air through neutral buffered potassium iodide solution. The effect of the ozone is to oxidise iodide to iodine, I_2:

$$O_3 + 2H^+ + 2I^- \rightarrow I_2 + H_2O + O_2 \tag{4.22}$$

The iodine reacts further with potassium iodide to form the strongly absorbing triiodide ion, I_3^-, which is determined by measuring its absorbance at 352 nm:

$$I_2 + KI \rightarrow KI_3 \tag{4.23}$$

The method is calibrated by making up standard solutions of iodine in potassium iodide and measurement of their absorbance to produce a calibration curve.

Analysis of atmospheric ozone may be more reliably carried out by measurement of its UV absorption at 254 nm using a long-path cell to enhance absorption (cf. equation 4.21). This is highly accurate and is a standard reference method for ozone in air.

Standard methods for analysis of many anions such as sulphate, nitrate and chloride are based upon UV–visible absorption methods. The rather recent advent of ion chromatography (Section 4.5), which offers advantages in terms of sensitivity, specificity and speed of analysis, provides a very valuable alternative procedure.

Atomic absorption spectroscopy

The electronic absorption spectra of atoms, unlike molecules, take the form of discrete lines of small bandwidth in the gas phase. These absorption lines are very narrow (less than 0.1 nm) and are characteristic of the analyte element. Thus by converting an element into the gas phase atoms, measurement of absorption of light at one of these characteristic wavelengths may be used to determine its concentration within a sample.

The instrumental set-up for atomic absorption spectroscopy is shown in Figure 4.13. It bears much similarity to a UV–visible spectrometer (Figure 4.11) and indeed operates within the same wavelength range. The main differences are two-fold. Firstly, the wide spectrum light source of the UV–visible spectrometer is replaced with a hollow cathode lamp. This lamp has a cathode coated with the analyte metal, and emits the characteristic resonance wavelengths of that metal at a very high intensity. Secondly, the sample cell is replaced by a flame, usually of air/acetylene or nitrous oxide/acetylene, into which a solution of sample is introduced as a fine spray. The effect of the flame is to volatilise the analyte and to dissociate and convert its ions into neutral atoms, which act as light absorbers. The preferred absorption wavelength is selected by the monochromator. The light absorption is calibrated by the use of standard solutions of the analyte metal. The beam chopper is used to pulse the light from the lamp, and the phase-sensitive detector is synchronised with this pulsing so as to respond only to signals which pulse at the frequency of this source. By this means unwanted stray light is excluded and the signal to noise ratio, and hence the sensitivity of the instrument, are enhanced. A different means by which sensitivity can be greatly improved is substitution of the flame by an electrically heated graphite rod or furnace. In a typical system, the sample solution ($10\text{--}100\,\mu l$) is injected into a hollow graphite tube located concentrically about the light beam. Temperature programming of the graphite tube causes solvent evaporation and subsequent vaporisation of the analyte at temperatures of 1900–2500°C as free atoms into the optical beam. The limits of detection for flame and graphite atomiser systems are cited in Table 4.6.

Atomic absorption methods are limited by practical constraints to metallic and metalloid elements, such as those listed in Table 4.6. In all some sixty-five elements are amenable to analysis by this means, although sensitivities vary substantially.

Atomic absorption spectroscopy is widely used in environmental analysis of metals. Almost any type of environmental sample may be analysed; whilst fresh water samples may be amenable to direct intro-

Figure 4.13. Schematic diagram of a single beam atomic absorption spectrometer.

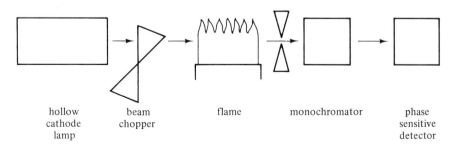

hollow	beam	flame	monochromator	phase
cathode	chopper			sensitive
lamp				detector

duction into the flame or graphite furnace, most other sample types require a digestion (Section 4.2.3) or pre-concentration/separation (Section 4.2.1) prior to analyses in order to put them into a suitable form for introduction into the instrument, and/or to eliminate other components of the sample which may influence the atomic absorption response (known as a matrix interference).

Infra-red absorption spectroscopy

The basic optical system of an infra-red spectrometer is very similar to that used for UV–visible work (see Figure 4.11), the main differences being that the source and detector are designed to operate in the infra-red (IR) region ($\sim$2.5–15 μm wavelength). At these wavelengths molecules absorb and emit radiation, not due to electronic

Table 4.6. *Typical atomic absorption detection limits*

Metal	Analytical line (nm)	Flame* (μg ml^{-1})	Graphite atomiser (pg)	Graphite atomiser (ng ml^{-1})§
Ag	328.1	0.0009	0.5	0.025
Al	309.3	0.30†	1	0.050
As	193.7	0.1	20	1.000
Ca	422.7	0.001	5	0.250
Cd	228.8	0.0005	0.3	0.015
Co	240.7	0.006	2	0.100
Cr	357.9	0.002	1	0.050
Cu	324.7	0.001	2	0.100
Fe	248.3	0.003	2	0.100
Hg	253.6	0.2	2000	100.0
K	766.5	0.002	2	0.100
Mg	285.2	0.00001	0.4	0.020
Mn	279.5	0.001	1	0.050
Mo	313.3	0.03	2	0.100
Na	589.0; 589.6	0.0002	<50	<2.500
Ni	323.0	0.004	10	0.500
Pb	283.3	0.01	5	0.250
Sb	217.6	0.03	20	1.000
Se	196.0	0.07	50	2.500
Sn	224.6	0.1†	20	1.000
V	318.3; 318.4; 318.5	0.04†	20	1.000
Zn	213.9	0.0008	0.1	0.005

* Double beam instrument; † Nitrous oxide – acetylene flame. Data cited for Perkin-Elmer spectrophotometer and heated graphite atomiser (data courtesy of Perkin-Elmer Ltd.). § For an estimated injection volume of 20 μl.

energy transitions, but to changes in vibrational energy within the molecule. Vibrational energy is associated with the stretching and bending of bonds in the molecule, and the absorption wavelengths are characteristic of the chemical bonds present.

A simple IR spectrum is shown in Figure 4.14. It may be seen that, although it is a band spectrum, the bands are relatively narrow in comparison to a UV–visible spectrum. Many of these absorption bands are highly characteristic of specific chemical functional groups within the molecule, and IR spectroscopy is a powerful tool for the organic chemist in the identification of unknown compounds.

Infra-red spectroscopy is not widely used in environmental analysis. One application is in the identification and quantification of silica (SiO_2) in workplace air and in sediments from its IR absorption. Another application, at the research level, involves the use of IR-absorption of the atmosphere over a one kilometre path length with Fourier transform signal processing, which allows identification and quantification of compounds such as ozone (O_3); peroxyacetyl nitrate ($CH_3C(O)OONO_2$); nitric acid (HNO_3); methanoic acid ($HCOOH$) and methanal ($HCHO$) at part per billion levels in polluted air. Also functional group analysis of dissolved organic substances (e.g. sea water) may be carried out by IR spectroscopy.

4.5   Separations

Analysis of gases in the atmosphere may be accomplished through introducing air samples directly into a suitable monitoring

Figure 4.14. IR absorption spectrum of cyclohexanone.

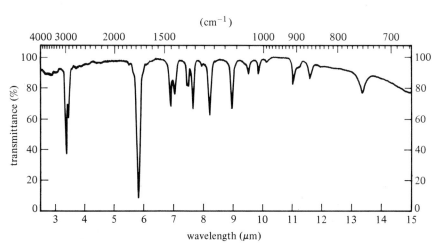

device; e.g. oxides of nitrogen (NO, NO_2) may be determined using a chemiluminescent analyser or CO_2 (and many organic vapours) using a detector employing infra-red absorption. Certain ions dissolved in natural waters can be determined directly using suitable equipment; hydrogen ion activity may be measured with a pH meter and electrode. Dissolved metals can be measured directly by atomic absorption spectrophotometry (AAS) or emission photometry. In such instances, the materials to be analysed occur in environmental media amenable to the analytical methodology and instrumental operating constraints. However, components of environmental interest are often associated with materials or substrates which are not directly suitable for introduction into instrumentation or particular wet chemical procedures. As a consequence it is necessary to transfer or extract the components of interest from the substrate in which they occur into a medium suitable for the type of analytical procedure to be used.

4.5.1 Extractions

Metals associated with particulate material, e.g. those adsorbed onto dust particles, may be transferred into solution by brief contact with dilute acids. Metals associated with the structure of soils may require prolonged boiling (digestion) with strong mineral acids in order to liberate them into solution prior to analysis. A series of extraction media may be used in a particular sequence to liberate metals associated with different soil components.

Organic and inorganic materials are often intimately associated with each other (see Section 5.4). For example, organic compounds entering the atmosphere (via natural processes such as fires or anthropogenic activities such as emissions from vehicles) may become adsorbed onto mineral material either in the atmosphere or through direct deposition onto soils. To extract the organics, the solid material is thoroughly mixed with solvent, in which the organic compounds dissolve. The resulting mixture of solutes may be rather difficult to analyse directly because of the many compounds likely to be present. Simple clean-up procedures are often used as preliminary steps in fractionating different components of the mixture prior to analysis. Polar organic compounds will dissolve in water. For example, phenols act as weak acids dissociating at high pH (>pH 12) thereby becoming completely soluble in water. This has been exploited for the extraction and separation of environmentally important chlorinated phenols. High pH does not enhance the water solubility of non-polar organic material or compounds lacking ionisable functional groups which remain associated with the organic solvent. Hence, thoroughly mixing water and organic solvent together causes the ionised species to enter the aqueous phase. The phenols may be extracted back into organic solvents if required by lowering the aqueous phase pH below pH 3.

Another type of extraction involves simply allowing a volatile component within a liquid to form an equilibrium between the liquid and vapour phase within an enclosed container. Vapour in the head space above the liquid surface can be withdrawn using a syringe and analysed by, for example, gas chromatography.

Organic solutes in aqueous environments are usually finely dispersed. They may be extracted by liquid partition as described previously for organics associated with particulate material. An alternative is to pass a large volume of water sample through a column packed with adsorbent capable of removing the solute from solution and retaining it. The adsorbed material, having been concentrated on the column packing in this way, can be desorbed either thermally or by solvent elution directly into an analytical instrument (gas or liquid chromatograph) (see also Section 4.5.2).

4.5.2 Chromatographic techniques

The term 'chromatography' is usually considered as covering the 'science of separations'. This science embraces techniques which enable components of a chemical mixture to be separated by virtue of differences in their physical and chemical properties. These differences control the extent of interaction between components of the mixture, carried in a mobile phase (a liquid or gas), with a solid matrix of material (the stationary phase). The stationary phase may simply be a sheet of paper (paper chromatography) or might consist of a finely divided solid, with or without an adsorbed or bonded coating, in the form of a thin layer on a glass or plastic sheet (thin layer chromatography), or packed in a column of glass or metal (as in liquid or gas chromatography).

Chromatography is a vast subject and is presented in many excellent texts which deal with theory and methodology (see the further reading suggestions at the end of this section). The following discussion is intended only as a guide to the major chromatographic techniques.

Gas chromatography (GC)

The essential components of a gas chromatograph are:

(a) a chromatographic column;

   (b) an oven and temperature programmer to heat the column in a controlled way;

   (c) a detector to respond to resolved components in a mixture as they leave the column;

   (d) a means of visualising the detector response, i.e. a chart recorded or computing integrator.

Separations achieved by GC rely on the volatility of components in a mixture and their affinity for the material coated onto the column packing. The mixture to be separated is introduced into the GC through a heated (usually) injection port and its vapour is carried through the column by an inert carrier gas, e.g. argon. Components of the mixture that are of low molecular weight, low boiling point (high volatility) and of low affinity for the column packing material, will reside in the GC column for the majority of the time in the gaseous phase and therefore move through the column to the detector quickly. High molecular weight components with high boiling points (low volatility) and of high affinity for the column packing will move slowly through the column as they remain predominantly in the stationary liquid phase. These are extremes in a spectrum of attributes which different compounds can exhibit. The key to successful separations is to match the appropriate column packing with the correct temperature programming and carrier gas flow rate for the type of compounds to be resolved. This sounds easy, but as any chromatographer knows it usually entails lengthy development work when one is attempting new types of separation.

   A major limitation of GC is that compounds which are to be separated must be volatile to some extent or else they will not elute from the column. In addition, as most separations require the column to be heated, compounds being introduced into the GC must be thermally stable or they will decompose. This may require the compounds to be derivatised, i.e. transformed, into a species which is both volatile and thermally stable. Methyl esters or trimethylsilyl ethers are common derivatives. Naturally not all compounds can be derivatised, and this leads to a major restriction as to the types of separations possible by GC.

   The most common GC detector is a flame ionisation detector (FID) which responds to the ions produced as the resolved compounds are burnt in a hydrogen air flame. There are other types of detector offering greater selectivity and sensitivity for specific groups of compounds. For example, the electron capture (EC) detector has been extensively used for compounds with electronegative functionality and has found particular application in the detection of the environmentally important polychlorinated biphenyls (PCBs). When coupled to a mass spectrometer (MS), the GC/MS system becomes a powerful analytical tool employing the separation offered by GC and the structural elucidation role of MS (Section 4.4.4).

Liquid chromatography (LC)

Whilst GC has made, and will continue to make, major contributions to analysis, not all compounds are amenable to the technique. Whilst many compounds with polar functional groups can be derivatised, it is estimated that only about 15% of all chemicals can exist in the vapour phase, hence the potential offered by separations accomplished entirely in the liquid phase. There are many LC techniques, e.g. paper chromatography (PC), thin layer chromatography (TLC), and high performance liquid chromatography (HPLC).

PC and TLC have been widely and successfully used for many applications because of the simple equipment required, ease of use and fairly good resolution of small quantities of sample. Solutions of sample or standard material to be separated are spotted onto an origin line drawn near one edge of either the PC paper or TLC plate. Components in each spot are eluted up (or occasionally down in the case of PC) the stationary phase using a suitable solvent system, the composition of which depends on the materials being separated. The ratio of the distance each component moves to the distance the solvent front moves from the origin is termed the Rf (retention factor). Comparison of sample Rf with those of known standard compounds enables identification of unknowns to be made. Visualisation of compounds on the paper or plate can be made in a number of ways depending on the compounds; e.g. spraying with agents forming a coloured product with specific compounds, charring with sulphuric acid, UV absorption (using plates coated with fluorescent indicator) or fluorescence. Ecologically important materials such as phytoallexins (naturally occurring compounds produced by some plants in response to infections by fungi) can be visualised on TLC plates by spraying with a suspension of Cladisporium spores. The fungus grows all over the plate except in places where the phytoallexins have moved to following solvent elution.

There is a major drawback to PC or TLC. Whilst separations may only take about one hour, particularly for TLC, measurement of the area or density of the visualised spots is time consuming and, if done visually by an operator, is very subjective. Precision better than 5% is seldom achieved.

HPLC has emerged as a particularly versatile and sensitive instrumental method for the separation and quantification of a wide range of compounds in the liquid phase (Figure 4.15). A basic HPLC system would consist of

(a) chromatographic column;
(b) pump to move the liquid mobile phase through the column at pressures which may exceed 5000 psi ($\sim$340 atm);
(c) a detector suitable for the type of compounds being analysed;
(d) chart recorder or computing integrator for data acquisition.

The great benefit offered by HPLC is that, providing the solute to be analysed is soluble in any one of a wide range of solvents (polar and non-polar), then there is a good prospect of achieving a separation and, if

Figure 4.15. Major components of HPLC and GC systems.

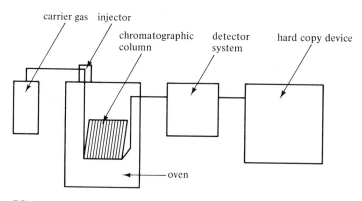

required, quantification. Early HPLC stationary phases were composed of silica or alumina using mobile phases of non-polar solvents such as hexane. This is referred to as normal phase chromatography, i.e. a polar stationary phase and non-polar mobile phase. Today, the vast majority of separations are achieved in reverse phase, i.e. employing a stationary phase which is less polar than the mobile phase. Reverse phase stationary phases typically consist of silica spheres (usually 5 or $10 \mu m$ in diameter) onto which are chemically bonded non-polar hydrocarbon chains. The most widely used coating appears to be octadecylsilane (ODS) having a chain length of eighteen carbons. There are many other types of bonded phase packing suitable for a wide range of applications. The presence of silica in these packings requires that separations are conducted at pH <7. New types of packing, based on polystyrene/divinylbenzene macroporous copolymers, are enabling some reverse phase separations to be conducted up to pH 13. This has proved of considerable benefit for the chromatography of coloured materials which undergo transformations to colourless species at pH <7.

Mobile phases for reverse phase separations are typically composed of either methanol or acetonitrile in combination with water enabling phase polarity to range from essentially non-polar at 100% organic solvent to polar with a totally aqueous phase. Mobile phase composition can be kept constant (isocratic) or can be gradually changed (gradient) during the separation. This offers considerable flexibility to the system and enables mixtures of solutes with vastly different affinities for the stationary phase to be separated without long run times and the concomitant detrimental effect this has on chromatographic parameters such as peak width and resolution.

There are many types of detector designed for specific groups of compounds. A refractive index (RI) detector is a type of universal detector. Since all solutes have a refractive index, and providing it is not the same as that of the mobile phase, then virtually anything can be detected. The drawback is the low sensitivity of RI detectors and the inability to use gradient elution with them as the constant change in mobile phase composition causes the RI to continually change. More specific detectors employing UV–visible absorption or fluorescence emission at particular wavelengths are used widely for compounds with chromophores (light-absorbing properties). Some compounds lacking a chromophore can still be detected using these detectors by adding a chromophore after the compound has been separated (post column derivatisation). Indirect UV absorption is a technique frequently used for analysis of solutes lacking chromophores, such as anions. An ion-exchange column is used with a mobile phase that is weakly UV absorbing, e.g. a dilute phthalate solution. The detector responds to the mobile phase by registering a uniform absorption. As resolved anions enter the detector they displace some of the phthalate and register as a negative peak (see the following subsection on ion chromatography).

The latest type of detector has enabled major advances to be made in LC. Photodiode array detectors enable instantaneous scanning across UV–visible wavelengths (190–700 nm) as each resolved compound leaves the column. This has assisted identification of compounds in a mixture. In addition, scans can be conducted at the start, apex and tail of each peak. If the three spectra from each peak differ, then it is likely that the peak does not represent a single entity but is composed of a mixture of at least two compounds. This is of considerable significance to the analysis of complex extracts often encountered in ecological and environmental studies.

An example of an HPLC chromatogram derived from extraction of a natural material (wood) is shown in Figure 4.16 where a wood preservative is identified from its chromatogram.

Ion chromatography

This is a liquid chromatographic technique that is concerned with the separation and detection of low molecular weight ionic species. Separation is effected on ion-exchange columns and detection is carried out by a number of detectors which are listed below. Its progress followed the known route of improving the liquid ion-exchange chromatography, with inherently time consuming processes (of separation of amino acid ions taking 10–70 hours), into an HPLC technique applied to amino acids and then to other lower molecular weight ions.

The ion-exchange process is based upon the equilibria between the eluting ions in solution and ions of the same sign on the surface of an ion-exchange resin. The most common cationic exchange sites on a resin are the sulphonic acid group $-SO_3^-H^+$ (strong acid) and the carboxylic acid group $-COO^-H^+$ (weak acid). Cationic exchange sites may contain tertiary amine groups $-N(CH_3)_3^+OH^-$ (strong base) or primary amine groups $-NH_3^+OH^-$ (weak base). The exchange equilibria for cationic exchange can be described as follows:

$$x RSO_3^-H^+ + M^{x+} \rightleftharpoons (RSO_3^-)_x M^{x+} + x H^+$$
$$\quad\text{resin}\qquad\text{solution}\qquad\text{resin}\qquad\text{solution}$$

Similarly for anionic exchange equilibria we have:

$$x\text{RN(CH}_3)_3^+\text{OH}^- + \text{A}^{x-} \rightleftharpoons [\text{RN(CH}_3)_3^+]_x\text{A}^{x-} + x\text{OH}^-$$

      resin   solution   resin   solution

The equilibrium constant for cation M^+ should be as follows:

$$\frac{\{\text{RSO}_3^-\text{M}^+\}(\text{s})\{\text{H}^+\}(\text{aq})}{\{\text{RSO}_3^-\text{H}^+\}(\text{s})\{\text{M}^+\}(\text{aq})} = K_{ex}$$

Figure 4.16. HPLC chromatogram of components extracted from wood treated with a preservative (mercaptobenzathiozole – MBT). The MBT peak is shown with a retention time of 5.76 minutes. Other peaks are naturally occurring UV absorbing compounds found in wood material. Conditions used to achieve the separation were: column, ODS (C18 packing); mobile phase, 60% acetonitrile, 40% water pH 3.0; flow rate, 1 ml min^{-1}; detector, UV absorption at 300 nm.

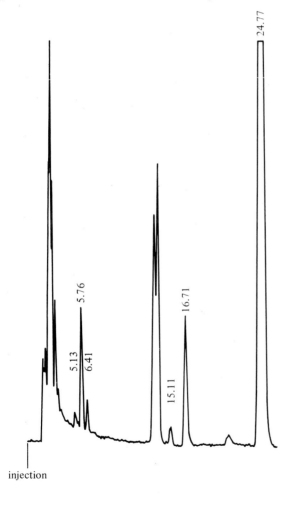

where s denotes solid resin. Rearranging gives

$$\frac{\{RSO_3^- M^+\}(s)}{\{M^+\}(aq)} = K_{ex} \frac{\{RSO_3^- H^+\}(s)}{\{H^+\}(aq)}$$

and because $\{RSO_3^- H^+\}(s)$ and $\{H^+\}(aq)$ are very large and not affected by shifts in the equilibrium, we can consider K_D, the partition coefficient, to be

$$K_D = \frac{C_s}{C_m} = \frac{\{RSO_3^- M^+\}(s)}{\{M^+\}(aq)}$$

i.e. C_s is the molar concentration of the analyte in the solid phase and C_m is its concentration in the aqueous phase.

There exist now a number of ion chromatography systems with different column configurations and detection systems. In the *suppressed ion chromatography* systems (e.g. Figure 4.17), ions are separated on a separator column that contains a low exchange capacity resin. A dilute eluent solution is used, and immediately following the anion-exchange column is the suppressor unit, which converts the eluent to a lower conductance compound (e.g. sodium carbonate to carbonic acid). This function lowers the background conductivity in comparison to the conductivity of the eluting ion, hence increasing sensitivity when a conductivity detector is used. Typical separations are shown in Figure 4.18.

Single column ion chromatography (SCIC) is based on the principle that an ion-exchange column of very low exchanging capacity (e.g. 0.005–0.100 meq g^{-1}) would need very low concentrations of eluents which inherently would have low conductivity when a conductivity detector is used. Typical eluents in this case are the sodium or potassium salts of benzoic or phthalic acids (for anion exchanging) and nitric acid or ethylenediammonium salt for monovalent or divalent cations, respectively.

In *ion-pair chromatography* a reversed phase or organic resin column is used along with an ionic modifier in the eluent. The ionic modifier acts as a movable ion-exchange site because the non-polar part of it is sorbed on the non-polar phase of the column and forms an ion pair with the sample ion with its ionic part. The sample ion is pushed down the column by a competing ion of the same charge in the eluent. This type of chromatography is much like ion-exchange chromatography with the modifier continually renewed by incorporation in the eluent.

The column materials that are used in anion exchange must have a very low exchange capacity (less than 0.1 meq g^{-1}) when ion chromatography is combined with conductivity detection. Resins of lower capacity permit the use of more dilute eluents, thus lowering the background conductance. Resins of higher capacity can be used with suppressed conductivity systems, but even so chromatographic and regeneration efficiencies dictate the use of low capacity resins. Silica-based ion exchangers with a quaternary ammonium group incorporated into the organic coating material have been used in anion chromatography. The separation efficiency of these silica based exchangers is quite good, although long term stability is inferior to organic ion-exchange

Figure 4.17. Schematic configuration of a suppressed conductivity ion chromatography system.

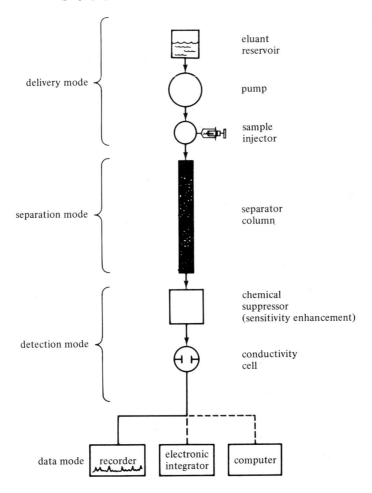

resins. Efficient anion-exchange columns can be prepared easily by coating porous polymer beads with a quaternary ammonium salt such as cetylpyridinium chloride.

Cation-exchange resins of low capacity can be prepared by sulphonating an organic polymer under mild conditions. Usually crosslinked polystyrene beads are sulphonated to effect surface sulphonation. A

Figure 4.18. Typical anion and cation separations by ion chromatography. (a) Separation of anions on ion-exchange column; eluent 0.0028 M $NaHCO_3$/0.0023 M Na_2CO_3; sample size 50 μl; suppressed conductivity detection. (b) Separation of alkaline earth ions on cation-exchange column; eluent 0.025 M phenylenediamine dihydrochloride/0.0025 M HCl; sample size 100 μl; suppressed conductivity detection.

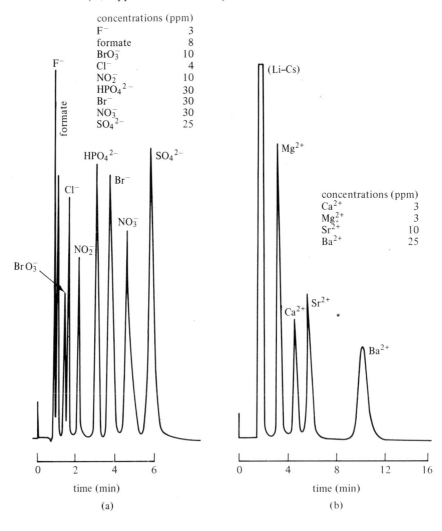

concentrations (ppm)

F^-	3
formate	8
BrO_3^-	10
Cl^-	4
NO_2^-	10
HPO_4^{2-}	30
Br^-	30
NO_3^-	30
SO_4^{2-}	25

concentrations (ppm)

Ca^{2+}	3
Mg^{2+}	3
Sr^{2+}	10
Ba^{2+}	25

time (min)

(a)

time (min)

(b)

reversed phase column coated with an alkyl sulphonate of higher molecular weight has been used for cation-exchange chromatography, but it is quite as effective to use ion pair chromatography with a modifier such as octane sulphonate in the eluent.

Detection systems in ion chromatography have been divided in three main categories:

(a) *Conductivity detectors*. These are used with suppressed conductivity and SCIC systems. Existing detectors will detect a conductivity change of one part per 30 000. If the whole system and not only the detector is under close temperature control, noise and baseline drift are improved. Coupled to a suppressor unit they are the most sensitive but delicate systems in ion chromatography.

(b) *Spectrophotometric detectors*. Direct photometric detection of nitrate, iodide and other ions that absorb in the UV region is possible with today's detectors. A more universal detection technique is to select an eluent that absorbs or fluoresces strongly in the UV or visible spectral region and at wavelengths at which sample ions do not absorb. During the ion-exchange process the ions elute after they have been displaced by eluent ions. This lowers the concentration of the eluent ions at the detector, producing a negative peak because of a decrease in absorbance. The phthalate or sulphobenzoate ions have been used successfully in the indirect photometric ion chromatography (IPIC). Metal ions and anions can also be detected after ion exchanging on a column, by post column derivatisation (i.e. reaction) processes using chromophores that absorb in the UV or visible spectral region.

(c) *Electrochemical detectors*. If they are of small volume (5–20 μl), these are cheap and very sensitive. Many types of organic ions, however, are adsorbed on noble metal electrodes, resulting in loss of sensitivity. Recently a pulsed amperometric detector has been developed (PAD) that makes constructive use of the adsorption of organic compounds on a noble metal electrode. The PAD detects organic molecules and radicals using the faradaic signal that results from their oxidative desorption. Following the detection process, a second potential is used for oxidative cleaning of the electrode surface and a third is employed to reduce the surface oxide on the electrode (see

Figure 4.19). The detection of amino acids by the PAD system is also possible.

Ion chromatography has been an extremely useful tool in analyses of acid rain samples and atmospheric acidity. Whole samples can be analysed for all the free anions or all the free cations. Recent attempts on the simultaneous detection of both inorganic anions and cations in one system have not been very successful.

Hybrid systems

Novel combinations of instrumentation have been employed for the analysis of particular compounds where chromatography is required to separate the components within a sample and a highly specific detector is used to respond only to particular entities. Organo-lead compounds (including tetramethyl and tetraethyl lead) added to petrol as anti-knock agents, have been determined using a combination of GC and AAS. These compounds are volatile at room temperatures and can be extracted from the atmosphere by passing air through a short column packed with an adsorbent. They can be thermally desorbed into a GC column and separated. The carrier gas leaving the GC column is diverted into the flame of an AAS set up to determine lead. As the resolved organo-lead compounds leave the GC and enter the AAS flame they are burnt, liberating lead which is detected by the AAS. The great advantage offered by the system is the specificity of the detector which responds only to lead. A simple chromatogram is therefore

Figure 4.19. Wave form used in a pulsed amperometric detector for organic ions separated by ion chromatography.

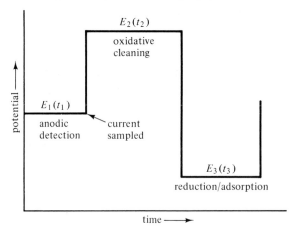

produced containing only organo-lead compounds; other materials present in the original air sample are not seen.

Worked examples

(1) If a sample in a 1 cm path cell with concentration of 0.5×10^{-3} M shows an absorbance of 0.265 absorbance units at 595 nm (λ_{max}) then its absorptivity or extinction coefficient is found as follows:

$$A = \varepsilon c l \text{ (Beer's law)}$$

$$\varepsilon = \frac{A}{cl} = \frac{0.265}{0.5 \times 10^{-3} \times 1 \times 10^{-2}}$$

$$\varepsilon = 53\,000 \text{ M}^{-1} \text{ m}^{-1}$$

(2) After passing 100 ml of 0.005 M Cs^+ solution through an ion-exchange resin (in its H^+ form of 2 meq g^{-1} capacity) the concentration of Cs^+ in solution was found to be 5 ppm.

The partition coefficient for the Cs^+ ion on the resin used is then calculated as follows:

$$K_D = \frac{(Cs^+)(s)}{(Cs^+)(aq)}$$

$$(Cs^+)(aq) = 5 \text{ mg l}^{-1} = \frac{5 \times 10^{-3} \text{ g l}^{-1}}{133 \text{ g}} = 0.0376 \times 10^{-3} \text{ M}$$

$$K_D = \frac{(0.005 - 0.0376 \times 10^{-3})}{(0.0376 \times 10^{-3})} = 132$$

(3) The determination of mercury in aqueous samples is carried out by cold vapour atomic absorption spectrometry. The sample is reduced with $NaBH_4$ in a reduction vessel and the resultant Hg^0 vapour is swept through a cylindrical cell positioned on the light path of a mercury lamp of the spectrometer. Analysis of two different samples gave mean absorbance values of 0.100 and 0.060. Standards of mercury in aqueous solutions gave the following results:

Conc. (ng ml^{-1})	20	40	60	80	100
Absorbance	0.03	0.05	0.08	0.11	0.13

After plotting a calibration graph (Figure 4.20) it was found that the samples contained 75 and 45 ng ml^{-1} of mercury, respectively.

Questions

(1) Define and elaborate on the following terms:
(a) qualitative analysis; (b) quantitative analysis;
(c) sensitivity; and (d) limit of detection.

(2) Define the terms 'accuracy' and 'precision'. Demonstrate with examples how an analytical method could be (a) accurate and precise, (b) accurate but imprecise, (c) inaccurate and precise, (d) inaccurate and imprecise.

(3) Which are the essential stages of an analysis? Elaborate.

(4) List some of the analytical techniques available today for environmental chemistry.

(5) Define the terms 'arithmetic mean', 'standard deviation' and 'rejection quotient'.

(6) What are the characteristics of a 'good indicator' in titrimetric analyses?

(7) What are the basic principles of the electrochemical methods? Give examples of electroanalytical techniques.

(8) List and elaborate on the operational principles of two X-ray analytical methods.

Figure 4.20.

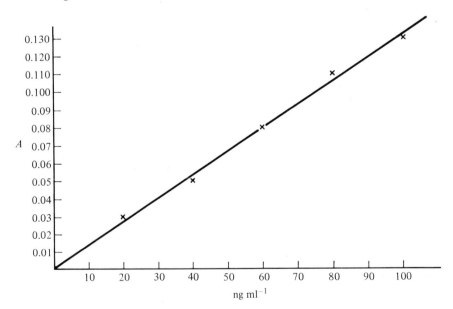

ng ml^{-1}

(9) How does a mass spectrometer scan a spectrum for identification of ions with different m/e? On what does the ion current for each peak in the spectrum of a monatomic element depend, and how does this help in the identification of an element in a complex matrix?

(10) Explain, with examples, how absorption spectroscopy can be used in environmental analysis.

(11) What is the prerequisite for the operation of an atomic absorption technique in contrast to molecular absorption techniques and how is this achieved?

(12) Summarise the different extraction procedures available for environmental samples.

(13) Elaborate on the essential components of gas chromatography.

(14) What are the most common stationary phases and detectors in HPLC and ion chromatography?

(15) A sample of river water (50 ml) is titrated with 0.100 M HCl, consuming 3.2 ml at the methyl orange equivalence point. What is the total alkalinity in units of mg l^{-1} ($CaCO_3$)?

References

Dean, R.B. & Dixon, W.J. (1951). *Analyt. Chem.* **23**, 636–8.
Hamilton-Taylor, J. (1983). *Environ. Technol. Lett.* **4**, 115–122.

Further reading

Braun, R.D. (1982). *Introduction to Chemical Analysis*. New York: McGraw-Hill.
Christian, G.D. & O'Reilly, J.E. (1986). *Instrumental Analysis*, 2nd edn. Boston: Allyn and Bacon.
Ewing, G.W. (1987). *Instrumental Methods of Analysis*, 5th edn. New York: McGraw-Hill.
Skoog, D.A. & West, D.M. (1986). *Analytical Chemistry: An Introduction*, 4th edn. Philadelphia: Saunders.

5

Case studies

5.1 Air chemistry

5.1.1 Reactive intermediates

The concept of reactive intermediates as short-lived highly reactive chemical species was introduced in Chapter 2. It is not, however, self evident that reactive intermediates, although in many cases present in the air at concentrations only of the order of *ca.* 10^{-7} ppm or less, play an enormously important role in atmospheric chemistry. Indeed their concentrations are so low that few, if any, direct measurements of their atmospheric concentrations have been made, and our knowledge of their concentrations is based partly upon calculation and inference.

Hydroxyl radical

This is the most important reactive intermediate in atmospheric chemistry. Its formation is dependent upon photochemical processes, and hence its concentration shows a pronounced diurnal variation exemplified by Figure 5.1.

There are many formation mechanisms for hydroxyl. These include photolysis of nitrous acid and hydrogen peroxide, and reaction of singlet state (excited) atomic oxygen, designated $O(^1D)$, (originating from ozone photolysis) with water vapour:

$$HNO_2 + h\nu \rightarrow OH + NO \tag{5.1}$$

$$H_2O_2 + h\nu \rightarrow 2OH \tag{5.2}$$

$$O_3 + h\nu \rightarrow O(^1D) + O_2 \tag{5.3}$$

$$O(^1D) + H_2O \rightarrow 2OH \tag{5.4}$$

There are also other routes of formation via the hydroperoxy radical (see later).

The hydroxyl radical reacts with many atmospheric gases, both natural and pollutant, in potentially important ways. For example, in a dry atmosphere, the most effective route of sulphur dioxide oxidation is by OH:

$$SO_2 + OH \rightarrow HOSO_2 \tag{5.5}$$

The $HOSO_2$ formed is converted to H_2SO_4, by reaction with O_2 and H_2O.

As mentioned in Chapter 2, one of the two important routes of nitric acid formation in air is from reaction of NO_2 with hydroxyl:

$$NO_2 + OH \rightarrow HNO_3 \tag{5.6}$$

Figure 5.1. Calculated concentrations of free radical species in the atmosphere as a function of season and time of day.

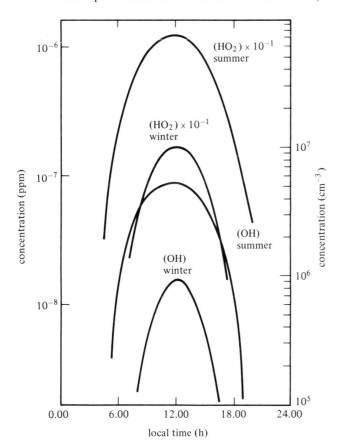

The hydroxyl radical also plays a major role in hydrocarbon oxidation in photochemical smog, and this will be exemplified later.

Peroxy radicals

The most important of these is the hydroperoxy radical HO_2. The estimated diurnal variation in the concentration of this radical is shown in Figure 5.1. Also important are the alkylperoxy radicals such as CH_3O_2 and $C_2H_5O_2$.

There are many routes of formation of the radicals. Processes leading to production of hydrogen atoms cause ultimately the formation of HO_2. These are, for example, photolysis of formaldehyde, and reaction of carbon monoxide with hydroxyl:

$$CH_2O + h\nu \rightarrow CHO + H \qquad (5.7)$$

$$CO + OH \rightarrow CO_2 + H \qquad (5.8)$$

$$H + O_2 \rightarrow HO_2 \qquad (5.9)$$

Alkylperoxy radicals are formed in a similar manner when alkyl radicals are generated, for example by oxidative cleavage of an alkene by atomic oxygen:

$$\overset{H}{\underset{H}{>}}C=C\overset{H}{\underset{H}{<}} + O \rightarrow CH_3 + CHO \qquad (5.10)$$

$$CH_3 + O_2 \rightarrow CH_3O_2 \qquad (5.11)$$

Peroxy radicals play a particularly important role in conversion of NO to NO_2 in air, the significance of which is explained later:

$$CH_3O_2 + NO \rightarrow CH_3O + NO_2 \qquad (5.12)$$

$$HO_2 + NO \rightarrow OH + NO_2 \qquad (5.13)$$

Note that in reaction (5.13) the HO_2 in reacting with NO has caused formation of both the highly reactive OH, and of photolysable NO_2.

5.1.2 Photochemical smog formation

Photochemical smog is a complex pollutant mixture generated when hydrocarbons and oxides of nitrogen interact in strong sunlight to cause formation of important and highly noxious secondary products.

The major products of the smog are:

(a) ozone, formed in concentrations up to *ca.* 0.4 ppm and toxic to humans and plants at this level;

(b) peroxyacyl nitrates. The most important of this group of compounds is peroxyacetyl nitrate (PAN), formed in concentrations up to *ca.* 0.05 ppm.

$$CH_3-C-O-O-NO_2 \qquad (PAN)$$
$$\overset{\|}{O}$$

The human toxicity of this compound has not been extensively studied but it is known to be toxic to plants.

(c) aldehydes, such as formaldehyde (HCHO), acetaldehyde (CH_3CHO) and acrolein ($CH_2{=}CH{-}CHO$). These cause odours and eye irritation;

(d) alkyl nitrates, such as methyl nitrate (CH_3ONO_2);

(e) aerosol – in other words airborne particles, which are responsible for the visibility reduction of the smog. The aerosol material is primarily ammonium sulphate, $(NH_4)_2SO_4$, and ammonium nitrate, NH_4NO_3, formed by neutralisation of sulphuric acid and nitric acid by atmospheric ammonia, itself released into the atmosphere mainly from decomposition of animal urine. Formation of the parent acids by photochemical mechanisms is much favoured in the smog. These are then neutralised:

$$H_2SO_4 + 2NH_3 \rightarrow (NH_4)_2SO_4 \qquad (5.14)$$

$$HNO_3 + NH_3 \rightarrow NH_4NO_3 \qquad (5.15)$$

Typical diurnal profiles of the major components of the smog are shown in Figure 5.2 and are taken from observations in the severely polluted Los Angeles area.

The typical sequence of chemical reactions in smog formation will be exemplified by a few of the processes. The chemistry is in fact enormously complex, particularly since a very wide range of hydrocarbons can take part in the reactions.

The smog is typically initiated by formation of atomic oxygen by photolysis of nitrogen dioxide:

$$NO_2 + h\nu \rightarrow NO + O \qquad (5.16)$$

As indicated in Chapter 2, the formation of oxygen atoms by this mechanism leads to ozone formation, which in turn causes reformation of NO_2:

$$O + O_2 + M \rightarrow O_3 + M \tag{5.17}$$

$$O_3 + NO \rightarrow NO_2 + O_2 \tag{5.18}$$

Small steady state concentrations of highly reactive oxygen atoms and ozone exist, and it is these which attack hydrocarbons. Alkenes are highly reactive, and typical processes are exemplified by attack on propene (reaction 5.19), which involves hydrogen atom transfer within the molecule after oxygen attack, and on butene (reaction 5.20)

$$CH_3CH{=}CH_2 + O \rightarrow C_2H_5 + CHO \tag{5.19}$$

$$CH_3CH{=}CHCH_3 + O_3 \rightarrow CH_3CHO_2 + CH_3CHO \tag{5.20}$$

CH_3CHO (ethanal or acetaldehyde) is the stable product, which may, however, undergo further reactions (see reaction 5.25). The reaction chain may then be propagated, for example:

$$C_2H_5 + O_2 \rightarrow C_2H_5O_2 \tag{5.21}$$

$$C_2H_5O_2 + NO \rightarrow C_2H_5O + NO_2 \tag{5.22}$$

$$C_2H_5O + O_2 \rightarrow CH_3CHO + HO_2 \tag{5.23}$$

$$HO_2 + NO \rightarrow NO_2 + OH \tag{5.13}$$

Now, both reactions (5.22) and (5.13) have converted NO to NO_2 without consumption of ozone. Photolysis of NO_2, followed by reaction (5.17) causes further formation of ozone. Then, if reactions (5.22) and (5.13) keep the concentration of NO low, reaction (5.18) is relatively

Figure 5.2. Typical diurnal profiles of major pollutants in a photochemical smog.

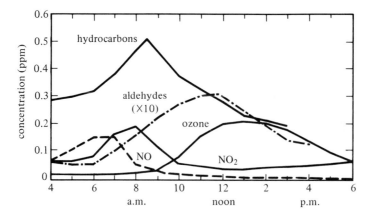

unimportant and high concentrations of ozone build up, typical of the photochemical smog.

Until rather recently, carbon monoxide was thought of as unreactive in the atmosphere, but it can in fact serve to enhance smog formation. By the sequence of reactions below, NO is converted to photolysable NO_2 without consumption of ozone, or of a reactive intermediate:

$$OH + CO \rightarrow CO_2 + H \qquad (5.24)$$

$$H + O_2 \rightarrow HO_2 \qquad (5.9)$$

$$HO_2 + NO \rightarrow NO_2 + OH \qquad (5.13)$$

Once formation of OH has commenced, the attack on hydrocarbons is considerably enhanced, as evidenced by Figure 5.3 which is taken from smog chamber studies.

The following reaction sequence involving acetaldehyde and the ethyl radical is one mechanism by which PAN may be formed:

$$CH_3CHO + C_2H_5 \rightarrow CH_3CO + C_2H_6 \qquad (5.25)$$

$$CH_3CO + O_2 \rightarrow CH_3C(O)O_2 \qquad (5.26)$$

$$CH_3C(O)O_2 + NO_2 \rightarrow CH_3C(O)O_2NO_2 \ (PAN) \qquad (5.27)$$

Methyl nitrate is formed by a simple termination process:

$$CH_3O + NO_2 + M \rightarrow CH_3ONO_2 + M \qquad (5.28)$$

Photochemical smogs are observed in Western Europe, as well as in

Figure 5.3. Calculated rates of reaction of various species with *trans*-2-butene in a simulated photochemical smog.

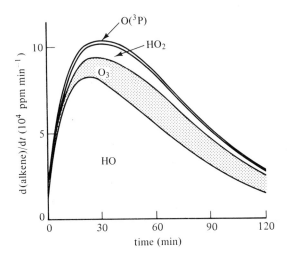

the USA. They tend to take a slightly different form and are distinctly less severe in Europe. Characteristically the European smog covers very large geographic areas simultaneously (hundreds or thousands of kilometres in scale) on the so-called 'synoptic' scale of an anticyclone. The ozone, characteristic of the smog, may be formed within hours of injections of the primary pollutants (hydrocarbons and oxides of nitrogen) but may then be transported over very large distances in an air mass. Thus Britain may experience ozone formed by precursor pollutants emitted in continental Western, or even Eastern, Europe.

5.1.3 Acid rain

This is a well publicised environmental problem. Pollutant emissions of the strong acids, nitric and sulphuric acid (HNO_3 and H_2SO_4), are slight, but atmospheric oxidation of NO_2 and SO_2 causes formation of these acids. They are removed from the air in rainfall, then known as acid rain.

Pure water in equilibrium with atmospheric carbon dioxide has a pH of approximately 5.6 due to release of hydrogen ions from ionisation of carbonic acid, H_2CO_3:

$$CO_2 + H_2O \rightleftharpoons H_2CO_3 \tag{5.29}$$

$$H_2CO_3 \rightleftharpoons HCO_3^- + H^+ \tag{5.30}$$

$$HCO_3^- \rightleftharpoons CO_3^{2-} + H^+ \tag{5.31}$$

Pollution by strong acids, however, causes far greater acidification, and pH values of 4 and below may be observed. (N.B. Since pH is on a logarithmic scale, a decrease of one unit represents a ten-fold increase in $\{H^+\}$.)

Sulphur dioxide is oxidised in the atmosphere by a number of mechanisms. It has two absorption bands for light of tropospheric (low atmosphere) wavelengths: a weak absorption at *ca.* 384 nm, giving rise to excited triplet state SO_2, and a strong absorption at *ca.* 294 nm, a wavelength of very low intensity in the low atmosphere, which causes formation of a higher energy excited singlet state. The following sequence of reactions is then possible:

$$SO_2 + h\nu \rightarrow {}^*SO_2 \tag{5.32}$$

$${}^*SO_2 + O_2 \rightarrow SO_3 + O \tag{5.33}$$

$$SO_2 + O \rightarrow SO_3 \tag{5.34}$$

$$SO_3 + H_2O \rightarrow H_2SO_4 \tag{5.35}$$

These reactions are in fact insufficiently rapid to explain the observed rates of SO_2 oxidation in the atmosphere (approximately 1–5% h^{-1}). It is probable that the major homogeneous (gas phase) mechanisms involve oxidation by reactive intermediates such as hydroxyl and peroxy radicals (see Section 5.1.1).

Reactions in cloud-water droplets appear to be important in wetter climates. When dissolved in water, SO_2 is in equilibrium with sulphite (SO_3^{2-}) and bisulphite (HSO_3^-) ions:

$$(SO_2)(g) + H_2O \rightleftharpoons (SO_2)(aq) \tag{5.36}$$

$$(SO_2)(aq) + H_2O \rightleftharpoons H_3O^+ + HSO_3^- \tag{5.37}$$

$$HSO_3^- + H_2O \rightleftharpoons H_3O^+ + SO_3^{2-} \tag{5.38}$$

The relative proportions of the dissolved species are dependent critically upon pH since hydrogen ions are involved in the latter two equilibria.

Sulphite may be oxidised by atmospheric oxygen, but this process is rather slow unless catalysed by transition metal ions $(Cu^{2+}; Fe^{3+})$. The overall reaction process is described by

$$2SO_3^{2-} + O_2 \rightarrow 2SO_4^{2-} \tag{5.39}$$

Thus sulphurous acid (H_2SO_3), a weak acid, is converted to H_2SO_4, a strong acid.

Other oxidation mechanisms are probably considerably faster. Oxidation of bisulphite by ozone is potentially significant, whilst reaction of sulphite with hydrogen peroxide is now believed to be very important:

$$HSO_3^- + O_3 \rightarrow HSO_4^- + O_2 \tag{5.40}$$

$$SO_3^{2-} + H_2O_2 \rightarrow SO_4^{2-} + H_2O \tag{5.41}$$

Oxidation of NO_2 to nitric acid is rather faster (*ca.* 10% h^{-1}) than for SO_2. There are two important mechanisms. The first, reaction with the hydroxyl radical (equation 5.6) has been discussed earlier, in Chapter 2. The alternative mechanism involves initial formation of the reactive intermediate NO_3, primarily by reaction of NO_2 with ozone:

$$NO_2 + O_3 \rightarrow NO_3 + O_2 \tag{5.42}$$

$$NO_3 + NO_2 \rightleftharpoons N_2O_5 \tag{5.43}$$

$$N_2O_5 + H_2O \rightarrow 2HNO_3 \tag{5.44}$$

Some hydrochloric acid gas, HCl, is emitted directly from coal combustion and refuse incineration. This may be an important contributor to acidity close to sources, whilst at greater range HNO_3 becomes important, and then at long distance (hundreds of kilometres) the rather slow oxidation of SO_2 becomes appreciable and H_2SO_4 is the main strong acid. All of these acids are neutralised to some degree by reaction with natural gaseous ammonia (equations 5.14 and 5.15).

5.1.4 Inorganic chemical composition of atmospheric particles and rain water

Chemical analysis of both rain water and of the water-soluble component of atmospheric aerosol particles normally reveals the presence of nine major ionic components. These are as follows:

(i) Na^+, Mg^{2+} and Cl^-, predominantly from sea spray;
(ii) K^+ and Ca^{2+}, predominantly soil-derived;
(iii) H^+ from strong acids;
(iv) NH_4^+ from ammonia neutralisation of strong acids;
(v) SO_4^{2-} and NO_3^- from oxidation of SO_2 and NO_2, respectively

There is occasionally also CO_3^{2-}, derived from soils and rocks, but this is expelled from rain water as CO_2 gas at lower pH values (see Sections 1.4.7 and 5.1.3) and hence is rarely present in this medium. It may be present in atmospheric particles but is rarely analysed due to interference from dissolution of atmospheric CO_2 in aqueous solutions derived from the particles for use in analysis.

Interesting properties are observed if concentrations of individual ions are expressed in terms of chemical equivalents. These are calculated as follows:

$$\text{concentration} = \text{concentration} \div \text{ionic weight} \times \text{ionic charge}$$
$$(\text{g eq } l^{-1}) \qquad\qquad (\text{g } l^{-1})$$

This is a form of expression as charge-equivalents, and thus for electroneutrality

$$\Sigma \text{ cations (g eq)} = \Sigma \text{ anions (g eq)}$$

Calculation of this charge balance equation is a very good indicator of the completeness and quality of analytical data. Some interesting relationships can also show up. If the ions Na^+, Mg^{2+} and Cl^- have a single common source, then

$$Na^+ + Mg^{2+} = Cl^-$$

when all concentrations are expressed in g eq. Also, if sulphate and nitrate are present solely as H_2SO_4, NH_4HSO_4, $(NH_4)_2SO_4$ and NH_4NO_3, and there are no other sources of H^+ and NH_4^+, then

$$NH_4^+ + H^+ = SO_4^{2-} + NO_3^- \text{ (all in g eq)}$$

Such charge balance relationships are commonly observed in rain water and atmospheric aerosols. Careful data analysis for samples collected in north-west England revealed excess NH_4^+ beyond that required by the above equation, and excess Cl^- beyond that needed to charge-neutralise $(Na^+ + Mg^{2+})$. The amounts were about equal, showing up the presence of NH_4Cl, formed from ammonia neutralisation of HCl from coal combustion and displacement from marine aerosol by reactions such as

$$2NaCl + H_2SO_4 \rightarrow Na_2SO_4 + 2HCl$$

$$NaCl + HNO_3 \rightarrow NaNO_3 + HCl$$

5.1.5 Stratospheric pollution

As mentioned in Chapter 2, stratospheric ozone plays an important role in absorption of potentially harmful solar ultra-violet radiation before it can penetrate to ground level.

The first cause of concern in stratospheric pollution was the supersonic transport aircraft which fly in the lower stratosphere. Their engines emit nitrogen oxide, which can interact in ozone formation processes as follows:

$$O_2 + h\nu \rightarrow 2O \qquad \lambda < 242 \text{ nm} \qquad (5.45)$$

$$O + O_2 + M \rightarrow O_3 + M \qquad (5.17)$$

$$NO + O_3 \rightarrow NO_2 + O_2 \qquad (5.18)$$

$$NO_2 + O \rightarrow NO + O_2 \qquad (5.46)$$

The sum of the two reactions involving nitrogen oxides (reactions 5.18 and 5.46) is

$$O + O_3 \rightarrow 2O_2 \qquad (5.47)$$

Thus NO is acting as a catalyst for ozone destruction. It cannot do so forever, as oxidation processes will convert it to HNO_3 which is eventually removed from the system.

Chronologically, the next cause for concern was the use of chloro-

fluorocarbons (freons). These compounds are used widely as refriger-
ants and as propellants in aerosol cans. They are very inert chemically,
and have no important tropospheric sink (i.e. removal process). They
do, however, diffuse through the air to the stratosphere on a long time
scale (tropospheric lifetime is estimated at 30–100 years), where photodissociation occurs with the short wavelength light available. The most
important compounds are $CFCl_3$ (freon 11) and CF_2Cl_2 (freon 12), the
former of which reacts as follows:

$$CFCl_3 + h\nu \rightarrow CFCl_2 + Cl \qquad (5.48)$$

$$Cl + O_3 \rightarrow ClO + O_2 \qquad (5.49)$$

$$ClO + O \rightarrow Cl + O_2 \qquad (5.50)$$

Thus in direct analogy with NO, the chlorine atom, Cl, is acting as a
catalyst for ozone decomposition.

Thirdly, there is concern that increased use of nitrate fertilisers will
lead to enhanced natural release of nitrous oxide, N_2O, from the soil.
Like the chlorofluorocarbons, N_2O is not effectively removed in the
troposphere, but in the stratosphere is photodissociated to give NO:

$$N_2O + h\nu \rightarrow N + NO \qquad \lambda < 250\,nm \qquad (5.51)$$

Regrettably, there is presently insufficiently detailed knowledge of
the chemistry and meteorology of the stratosphere to enable a very
reliable prediction of the effects of these pollutants upon the concentrations of ozone. However, a firm connection between chlorofluorocarbons and Antarctic ozone depletion (the 'ozone hole') has been
established.

5.2   Fresh waters

5.2.1 Composition of natural fresh waters

The hydrological cycle plays a fundamental role in the geochemical cycling of elements. The ocean constitutes by far the largest
reservoir of water and the major concepts in marine chemistry are
outlined in Section 5.3. While fresh waters in rivers, lakes and the
atmosphere constitute only 0.003% of the hydrosphere, they are
responsible for several of the features of the hydrological cycle. Rain
and snow scavenge marine-derived material from the atmosphere.
Water is an excellent solvent and, in the form of rain or soil solution,
initiates the weathering (i.e. disintegration) of continental rock

material. Rivers act as the major agent for the erosion and transport of continental material to the oceans. The composition of rain and river waters will be discussed here while weathering processes will be considered in Section 5.2.2.

Rain water

Water which precipitates as rain and snow is largely derived from the sea. The residence time of water in the atmosphere is of the order of 11 days compared to 1 year on the continents in soil, rivers, lakes, etc., and about 3500 years in the ocean. Aerosols act as the nucleation centre for water vapour condensation. These fine atmospheric particles, predominantly in the size range 0.1 to 20 μm diameter, may be derived from the continents or the oceans. Continental material comprises aluminosilicates and refractory (i.e. resistant to weathering) oxides. Marine aerosols consist of highly soluble chloride and sulphate salts. These form when breaking bubbles inject droplets of sea water into the atmosphere; subsequent dehydration leaves particles with a composition resembling sea salt. An exception is that sodium can be enriched with respect to chlorine possibly due to losses of the latter as volatile HCl.

The composition of 'average' rain water is presented in Table 5.1.

Table 5.1. *'Average' concentrations of the major constituents dissolved in rain and river water*

Constituent	Concentrations (mg dm^{-3})	
	Rain water	River water
Na^+	1.98	6.3
K^+	0.30	2.3
Mg^{2+}	0.27	4.1
Ca^{2+}	0.09	15
Fe		0.67
Al		0.01
Cl^-	3.79	7.8
SO_4^{2-}	0.58	11.2
HCO_3^-	0.12	58.4
SiO_2		13.1
pH	5.7	

From Garrels & MacKenzie (1971).

Marked variations can occur due to enhanced contribution from continental aerosols or anthropogenic inputs. As discussed in Section 5.1.3 emission of SO_2 and NO_x may greatly increase the acidity of rain water. In general, the salt concentration in rain water decreases towards the continental interiors, i.e. with distance from marine source, when anthropogenic inputs are negligible.

River water

The flux of material from the continents to the oceans is dominated by riverine transport. The 'average' concentrations of the principal constituents dissolved in river water are given in Table 5.1. Wide regional variations exist and three main contributions determine the composition and concentration of material in river water: (1) the salts in rain water; (2) the weathering and erosion of continental material in the catchment area; (3) anthropogenic discharges. Weathering and erosion processes are discussed in Section 5.2.2.

The predominant cations in the surface waters of the world are Na^+ (sea water) and Ca^{2+} (fresh water). A plot of total dissolved solids

Table 5.2. *A comparison of the concentration of major elements in 'average' riverine particulate material and surficial rocks*

	Concentrations (g kg^{-1})	
Element	Riverine particulate material	Surficial rocks
Al	94.0	69.3
Ca	21.5	45.0
Fe	48.0	35.9
K	20.0	24.4
Mg	11.8	16.4
Mn	1.1	0.7
Na	7.1	14.2
P	1.2	0.6
Si	285.0	275.0
Ti	5.6	3.8

From Martin & Meybeck (1979).

versus the weight ratio $Na^+/(Na^+ + Ca^{2+})$ is presented in Figure 5.4. Two diagonal lines emerge indicating that the composition of the world's surface waters can be explained in terms of three mechanisms (Figure 5.5). The *precipitation dominance* regime occurs for rivers with low levels of dissolved solids which flow from catchment areas

Figure 5.4. Variation of the weight ratio Na/(Na + Ca) as a function of the total dissolved salts for several surface waters. (From Gibbs, 1970.)

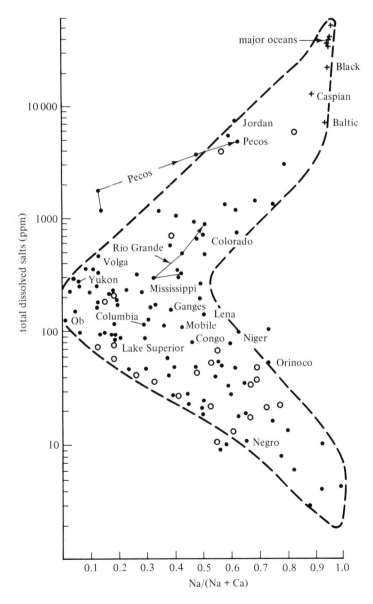

exhibiting little weathering or erosion. Accordingly, the salt matrix is derived from the precipitation which in turn reflects a marine origin. The river salt therefore is sodium rich. The *rock dominance* regime is characterised by intermediate total dissolved solids. The enhanced salt concentration is derived from the weathering of rock material. Calcium is the major cation, particularly for rivers draining areas containing carbonate minerals. The third regime, *evaporation/crystallisation*, exhibits still greater total dissolved solid loads. This further enhance-

Figure 5.5. Diagrammatic representation of processes controlling the chemistry of world surface waters. (From Gibbs, 1970.)

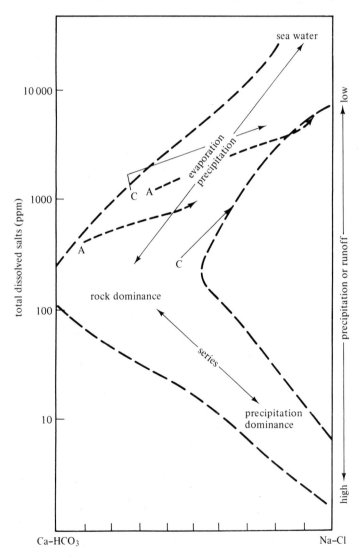

ment occurs by the evaporation of water, and hence this mechanism is confined to rivers in hot, arid regions. Sodium again becomes the predominant cation due to the precipitation of calcium carbonate.

The global riverine flux of dissolved solids to the ocean is 4.2×10^{12} kg y^{-1}. In contrast, rivers contribute about 18.3×10^{12} kg y^{-1} of particulate (suspended) solids. Environmental chemists differentiate dissolved and particulate fractions in natural waters by filtration using filters with a nominal pore size in the range 0.4–0.5 μm. The concentrations are compared for 'average' riverine particulate material (RPM) and surficial rocks for elements with RPM > 1 g kg^{-1} in Table 5.2. Elemental enrichment in RPM relative to surficial rocks is indicative of greater resistance to chemical weathering (aluminium, iron, titanium) or biological activity (phosphorus). Alternatively, elements susceptible to chemical weathering are depleted in RPM (calcium, sodium).

The discussion above centres on the total amount of the principal constituents dissolved and suspended in river water. This is adequate for estimation of riverine fluxes and establishment of the role of weathering processes. An understanding of the chemistry requires a greater appreciation of the *physico-chemical speciation*. This term refers to the physical (i.e. dissolved, colloidal, particulate) and chemical (i.e. free ionic, complexed, adsorbed) nature of the element. As indicated earlier, elements in natural waters are usually only size-differentiated by filtration into dissolved and particulate fractions. Chemical fractionation of the particulate material can be used to identify anthropogenic inputs associated with RPM. In Figure 5.6, the speciation of iron, manganese, zinc, copper, nickel, lead and cadmium in eighteen riverine sediments is illustrated. The fractions identified using the specified chemical extractants are: exchangeable ($CH_3CO_2NH_4$), associated with Mn–Fe oxide surface coatings (NH_2OH/HCl), organically associated (H_2O_2/HCl) and mineral lattice or resistant ($HF/HClO_4$). In unpolluted rivers such as the Great Ruaha and the Orinoco, the trace metals are predominantly associated with the resistant fraction and are derived by weathering mechanisms. In polluted rivers (Elbe, Somme), trace metals of anthropogenic origin occur in exchangeable fractions or in association with surface coatings of manganese and iron oxyhydroxides.

Lake water

The chemical composition of lake waters will naturally reflect the composition of their riverine sources. However, significant differences between river and lake waters may arise due to physical and biological factors. Considering firstly physical processes, the hydro-dynamic energy in the lake is considerably less than that of rivers. Consequently, RPM carried in suspension may be deposited upon

Figure 5.6. The speciation of trace metals in eighteen different river sediments, arranged according to their approximate geographic position from north to south. Most tropical rivers contained low cadmium levels, and no reliable data were obtained. For the Rio Magdalena and Orinoco River, insufficient material was available for determination of the 'exchangeable' (NH$_4$OAc) fraction, and this is contained in the hydroxylamine extract. (From Salomons and Forstner, 1980.)

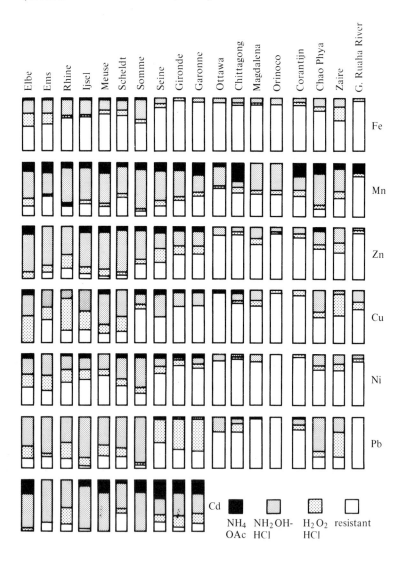

entering the lake. The steep topography of lake bottoms causes *sediment focusing* whereby sediments accumulate at a greater rate in the lake basins than in the peripheral shallow zone. In summer months, the high incident radiation to surface waters causes the temperature to rise. Thermal stratification may produce a two-layer system in which a warm, buoyant, surface layer overlies a deep, colder layer. Wind induced mixing during the late autumn and winter re-establishes isothermal conditions in the lake.

Thermal stratification in the summer may become established to such an extent that exchange of material across the *thermocline* (i.e. sharp temperature gradient which indicates a boundary between water masses of different temperature) is effectively prevented. Biological activity under such conditions can produce marked compositional gradients. In the surface layer, oxygen levels remain high due to equilibration with the atmosphere and photosynthetic activity. However, respiratory consumption of oxygen in the bottom waters may completely remove the oxygen producing an anoxic environment. Reducing conditions (i.e. lower Eh) are established. Metals will exist in lower oxidation states, i.e. Fe^{2+}, Mn^{2+} cf. Fe^{3+}, Mn^{4+} and NO_3^-, NO_2^- and SO_4^{2-} will be used successively as biochemical oxidants.

5.2.2 Weathering processes

Both the composition and concentration of riverborne material are determined by weathering and erosion. *Weathering* refers to the *in situ* disintegration of the parent rock material at or near the interface of the atmosphere and the lithosphere. *Erosion* may be similar but also implies transport of the material from the site of weathering. Two types of processes may be characterised. Physical (or mechanical) weathering involves the fragmentation of rock matrix with the proviso that the chemical integrity of the original material is retained. Alternatively, chemical weathering involves the interaction of parent rock with natural waters to produce weathered residues with compositions different to the original rock matrix. These two categories of weathering mechanisms are not mutually exclusive but both types depend upon the local environment. The type and extent of weathering depends upon the composition of the rock matrix, the physico-chemical characteristics of the natural waters and the climate. With regard to climate influences, weathering will be affected by the temperature and precipitation which in turn determine two further important weathering criteria, namely vegetation cover and soil formation. As a manifestation of climatic control, maximum chemical weathering occurs in equatorial regions.

Physical weathering involves the fragmentation of parent rock. While the surface area of exposed rock is greatly enhanced, the chemical composition remains unaltered. Several mechanisms are responsible for physical weathering. Cyclic heating and cooling of rocks due to diurnal temperature variations may cause fractures. This process is accelerated when water is present due to hydrolysis (i.e. chemical weathering). Frost shattering or heaving of rocks is important in environments subject to repeated freeze–thaw cycles. Fragmentation occurs due to specific volume changes of the water filling cracks and crevices in the rock. Fractures in rocks may be extended by root growth. Again, this process may be enhanced chemically due to the release of organic acids into the soil solution during plant growth. Some erosional processes contribute towards physical weathering. Glacial erosion involves plucking and abrasion. In the first instance, blocks of rock become engulfed in ice, and the steady downward flow of ice wrenches the material from the land surface. Such blocks become embedded in the lower surface of the ice sheet and abrade the exposed rock to produce glacial flour. Wind and water behave in a similar but much less efficient manner. Rock fragments borne in suspension bombard exposed parent rock causing abrasion and fragmentation. Unloading of rocks, caused by the erosion of overlying material, is accompanied by pressure reductions which in turn may open fissures and fractures originally established during the rock formation.

Chemical weathering results from the reaction of natural waters (containing dissolved gases and solids) with parent rock. The end result may be the partial (or complete) dissolution of the rock together with the formation of new mineral phases. Mechanisms of chemical weathering which will be outlined below are: dissolution, hydration, hydrolysis, carbonation and oxidation.

Dissolution

Rock material may contain several soluble constituents. Dissolution may be described as congruent (simple) or incongruent. During *congruent dissolution*, the cationic and anionic electrolytes are dissolved in proportion to their stoichiometric relationships in the solid phase. The dissolution of halite ($NaCl$), and quartz (SiO_2) exemplify congruent dissolution. Halite is readily soluble with a solubility of $\sim350\,g\,kg^{-1}$ of water at 25°C:

$$NaCl(s) \rightleftharpoons Na^+(aq) + Cl^-(aq)$$

Thus, equimolar amounts of Na^+ and Cl^- are released into solution. As H^+ is not involved, halite solubility is independent of pH. This contrasts with the dissolution of quartz (solubility about 6.6 g kg^{-1} at 25°C):

$$SiO_2(s) + 2H_2O \rightleftharpoons H_4SiO_4$$
$$\text{quartz} \qquad\qquad \text{silicic acid}$$

Silicic acid is a weak acid (pK = 9.8)

$$H_4SiO_4 \rightleftharpoons H^+ + H_3SiO_4^-$$

According to Le Chatelier's principle, as the pH of the solution rises, the reaction would be pushed to the right causing an increase in the solubility.

Incongruent dissolution involves the formation of a new mineral phase due to the differential leaching of ions from the original rock matrix. The dissolution of serpentine and orthoclase are examples:

$$\left\{ 4Mg_3Si_2O_5(OH)_4 + 12H^+ \rightarrow Mg_6Si_8O_{20}(OH)_4 + 6Mg^{2+} + 12H_2O \right.$$
$$\text{serpentine}$$

$$\left\{ 4KAlSi_3O_8 + 22H_2O \rightarrow Al_4Si_4O_{10}(OH)_8 + 8Si(OH)_4 + 4K^+ + 4OH^- \right.$$
$$\text{orthoclase} \qquad\qquad \text{kaolinite}$$

In the case of serpentine, only magnesium is extracted from the mineral. Incongruent dissolution of orthoclase leaches both silica and potassium. As H^+ or OH^- are involved in these examples, the dissolution is pH dependent and the solubility can thus be affected by microbial activity in the soil or sediment which produces CO_2 and hence carbonic acid.

Hydration and hydrolysis

Hydration and hydrolysis are important and closely related weathering mechanisms involving the interaction of water and the rock material. During *hydration*, water molecules are incorporated into the crystalline structure of a compound (i.e. water of crystallisation) but do not react with the compound. Weathering by hydration involves the surface adsorption of water molecules, such as the case for haematite, (Fe_2O_3):

$$Fe_2O_3(s) + 3H_2O \rightarrow Fe_2O_3 \cdot 3H_2O(s)$$

Hydrolysis involves the chemical reaction of water with the rock matrix and is manifest in the breaking of oxygen–hydrogen bonds. The simplest examples are the surface adsorption of H^+ and OH^- at appropriate Brønsted base and Lewis acid sites, respectively. The mechanism actually involves surface hydration followed by dissociation of the water molecule. Weathering of the rock matrix further requires the disruption of element–oxygen bonds within the crystal lattice. Therefore, the elements which form the most stable bonds with oxygen (i.e. silicon, aluminium and titanium) exhibit the greatest resistance to chemical weathering (i.e. quartz, gibbsite and rutile). Hydrolytic weathering processes may produce solutions which are either acidic or basic as exemplified by the reactions below. The hydrolysis of $Fe_2(SO_4)_3$ creates an acidic solution due to the dissociation of H_2SO_4:

$$Fe_2(SO_4)_3(s) + 6H_2O \rightarrow 2Fe(OH)_3(s) + 3H_2SO_4(aq)$$

$$3H_2SO_4(aq) \rightarrow 6H^+(aq) + 3SO_4^{2-}(aq)$$

A basic solution occurs following the hydrolysis of CaO:

$$CaO(s) + H_2O \rightarrow Ca(OH)_2(aq)$$

$$Ca(OH)_2(aq) \rightarrow Ca^{2+}(aq) + 2OH^-(aq)$$

Carbonation

Carbonation refers to the chemical weathering of rock material by reactions with carbon dioxide. In solution, CO_2 reacts with H_2O forming H_2CO_3. The subsequent dissociation of carbonic acid imparts acidity to the waters. Carbonation processes are a category of acid hydrolysis, comparable to hydrolysis mechanisms but with acidic waters and hence more reactive. Typical carbonation weathering processes for calcite and dolomite are:

$$CaCO_3(s) + CO_2(g) + H_2O \rightleftharpoons Ca^{2+}(aq) + 2HCO_3^-(aq)$$

$$CaMg(CO_3)_2(s) + 2CO_2(g) + 2H_2O \rightleftharpoons Ca^{2+}(aq) + Mg^{2+}(aq) + 4HCO_3^-(aq)$$

Carbonation is not necessarily restricted to carbonate minerals but can also affect silicates, as indicated for the weathering of forsterite and microcline:

$$Mg_2SiO_4(s) + 4CO_2(g) + 4H_2O \rightarrow 2Mg^{2+}(aq) + 4HCO_3^-(aq)$$
forsterite

$$+ H_4SiO_4(aq)$$

$$4KAlSi_3O_8(s) + 4CO_2(g) + 22H_2O \rightarrow Al_4Si_4O_{10}(OH)_8(s) + 4K^+(aq)$$
microcline   kaolinite

$$+ 4HCO_3^-(aq) + 8H_4SiO_4(aq)$$

The weathering of microcline to kaolinite is a further example of incongruent dissolution. Note that, while microcline and orthoclase have similar chemical composition, they do exhibit different crystal structures.

Oxidation

Oxidation, the removal of electron(s) from an element which can exhibit several oxidation states, is an important chemical weathering process. Minerals formed in anoxic (oxygen-free) waters possess elements in reduced forms. The exposure of such minerals to oxygenated waters results in dissolution followed by oxidation. Equilibrium may be finally attained with the precipitation of a new mineral phase. Consider the stepwise oxidative weathering of rhodochrosite ($MnCO_3$):

(i) dissolution

$$MnCO_3(s) \rightleftharpoons Mn^{2+}(aq) + CO_3^{2-}(aq)$$

or in the presence of dissolved CO_2

$$MnCO_3(s) + CO_2(g) + H_2O \rightleftharpoons Mn^{2+}(aq) + 2HCO_3^-(aq)$$

(ii) oxidative precipitation

$$2Mn^{2+}(aq) + 2H_2O + O_2(g) \rightleftharpoons 2MnO_2(s) + 4H^+(aq)$$

Chemical weathering by oxidation is especially important for elements which form insoluble sulphides. For instance, both iron and sulphur are oxidised in the weathering of pyrite (FeS_2) during a complex series of reactions:

$$2FeS_2(s) + 7O_2(g) + 2H_2O \rightarrow 2Fe^{2+}(aq) + 4SO_4^{2-}(aq) + 4H^+(aq)$$

$$4Fe^{2+}(aq) + O_2(g) + 4H^+(aq) \rightarrow 4Fe^{3+}(aq) + 2H_2O$$

$$Fe^{3+}(aq) + 3H_2O \rightarrow Fe(OH)_3(s) + 3H^+(aq)$$

$$FeS_2(s) + 14Fe^{3+}(aq) + 8H_2O \rightarrow 15Fe^{2+}(aq) + 2SO_4^{2-}(aq) + 16H^+(aq)$$

The major agencies of physical and chemical weathering have been outlined above. The rock material which will subsequently be transported to the sea will consist of material resistant to chemical weathering, components subjected to little or no chemical weathering at source and the weathering residues (i.e. the new minerals produced by incongruent dissolution processes), plus, of course, the material dissolved during the weathering processes.

5.2.3 Lead contamination in drinking water

An important type of fresh water to consider is drinking water. While several water quality criteria are regulated, of which microbiological control is undoubtedly the most important, we shall concentrate here on lead. Lead is toxic and has no known beneficial biochemical attributes. Its presence in tap water has attracted considerable recent interest as maximum admissible concentrations (MAC) have been decreased. The MAC for lead in tap water is now $50\,\mu g\,dm^{-3}$ in most countries. Low concentrations of lead, generally $<25\,\mu g\,dm^{-3}$, are found in surface and ground waters which are used as a source for drinking water supplies. These concentrations usually diminish further due to water treatment processes prior to distribution. Hence, lead levels which exceed the MAC are caused by contamination from lead components in the distribution system such as lead pipes and water storage tanks. In houses with copper pipes, contamination may be caused by lead-based solder joints.

Several factors influence the extent to which lead contamination may occur. The presence of lead pipes or solder joints alone does not constitute a threat to water quality. The physico-chemical characteristics of the water are important as a layer of scale can be deposited on the internal pipe surfaces. Depending upon the nature of the scale deposit, the scale may prevent or retard contamination due to dissolved lead but contribute to particulate lead levels. Regardless of the mechanism responsible, lead concentrations in tap water increase with contact time, an increase in the total area of exposed pipe surface and a decrease in the ratio of pipe volume:surface area. A number of corrosion mechanisms will be outlined.

Galvanic corrosion

Dissolved oxygen in drinking water can oxidise lead via the reaction:

$$2Pb + O_2 + 2H_2O \rightarrow 2Pb^{2+} + 4OH^-$$

Mixed metal contacts, such as the union of pipes of different metals or lead-based solder joints for copper pipes, may stimulate galvanic corrosion via the reaction:

$$Cu^{2+} + Pb(s) \rightarrow Cu(s) + Pb^{2+}$$

Copper oxidation occurs due to dissolved oxygen as shown above for lead. Downstream the copper may be adsorbed onto exposed lead pipe or solder joints. Consequent copper reduction corrodes the lead.

Solubility of Pb^{2+}

The *plumbosolvency* (i.e. susceptibility towards lead contamination) will be determined by the chemical composition of the water and the nature of the pipe scale in equilibrium with that water. The scale should consist of basic lead carbonate, $Pb_3(CO_3)_2(OH)_2$, at high pH and normal lead carbonate, $PbCO_3$, at low pH. As evident in Figure 5.7, the solubility of lead is greatest in acidic waters of low alkalinity. Lead solubility increases with temperature, organic complexation or adsorption onto particulate material. Remedial action (i.e. treatment at a water works prior to distribution) relies upon reducing the solubility of pipe scale deposit. For acidic, low alkalinity waters this may involve increasing the pH and the alkalinity. Phosphate dosing (Na_2HPO_4 at 1 mg P dm^{-3}) is used for hard waters; the solubility then being determined by $Pb_3(PO_4)_2$ or $Pb_5(PO_4)_3OH$.

Particulate lead

Several physical and chemical processes may cause tap water contamination with particulate lead. Pipe scale can be dislodged by vibrational effects such as water hammer or scoured from the pipes due to fluctuations in water pressure. Chemical inhomogeneities in the pipe scale may produce a friable (i.e. prone to crumbling or flaking) deposit. Variations in the physico-chemical characteristics of the drinking water (i.e. pH, alkalinity, dissolved oxygen concentration) may cause the scale to be either sloughed off or dissolved. Finally, the presence of particulate material can act as a sink for Pb^{2+}. Adsorption onto the particulate material prevents the attainment of equilibrium between the water and scale causing dissolution of the pipe deposit. It should be noted that the contamination of tap water by particulate lead cannot be predicted by thermodynamic models but must be determined empirically.

5.3   Sea water

5.3.1 Composition

The oceans represent a chemical system on a massive scale, covering 71% of the earth's surface and accounting for 77% of the water

Figure 5.7. The theoretical concentration as a function of pH for dissolved lead in equilibrium with basic and normal lead carbonate for (a) low alkalinity (10 mg dm^{-3} as $CaCO_3$) and (b) high alkalinity (250 mg dm^{-3} as $CaCO_3$) waters. (From Gregory & Jackson, 1983.)

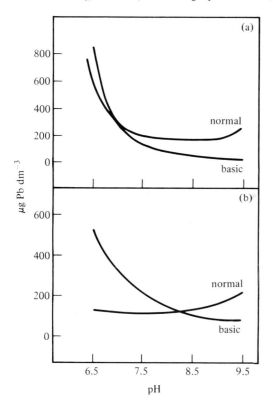

in the global hydrological cycle. The average depth of the ocean is 3.7 km but the water can be greater than 10 km in the deep-sea. Sea water is a solution of gases and solids containing both organic and inorganic material in suspension. This is a well-oxygenated solution buffered at about pH 8 containing all the elements, albeit in some instances in very dilute solution. In a thermodynamic sense the oceans form an open system as both energy and mass can be exchanged across the boundaries. It may be convenient in some circumstances to consider the ocean as a homogeneous system at equilibrium. This must at best be considered an extreme simplification but apparently holds true for the major constituents. Sea water is both physically and chemically dynamic in nature leading to heterogeneity within the system. Equilibrium may be attained for some reactions requiring geologic time scales. Short term perturbations may upset these equilibria. Of greater importance may be the effect of organisms. The concentration and distribution of several elements is controlled not chemically but biologically. Finally, while some parameters may exhibit relatively uniform distribution in the bulk solution of the ocean, atypical marine environments may exist at the boundaries.

The major elements in sea water (Table 5.3) are conventionally taken

Table 5.3. *The major elements in sea water*

Element	Chemical species	Concentration for $S = 35‰$		Ratio to chlorinity $(Cl = 19.374‰)$
		(mol dm^{-3})	(g kg^{-1})	
Na	Na^+	4.79×10^{-1}	10.77	5.56×10^{-1}
Mg	Mg^{2+}	5.44×10^{-2}	1.29	6.66×10^{-2}
Ca	Ca^{2+}	1.05×10^{-2}	0.4123	2.13×10^{-2}
K	K^+	1.05×10^{-2}	0.3991	2.06×10^{-2}
Sr	Sr^{2+}	9.51×10^{-5}	0.00814	4.20×10^{-4}
Cl	Cl^-	5.59×10^{-1}	19.353	9.99×10^{-1}
S	$SO_4^{2-}, NaSO_4^-$	2.89×10^{-2}	0.905	4.67×10^{-2}
C (inorganic)	HCO_3^-, CO_3^{2-}	2.35×10^{-3}	0.276	1.42×10^{-2}
Br	Br^-	8.62×10^{-4}	0.673	3.47×10^{-3}
B	$B(OH)_3, B(OH)_4^-$	4.21×10^{-4}	0.0445	2.30×10^{-3}
F	F^-, MgF^+	7.51×10^{-5}	0.00139	7.17×10^{-5}

Based on Dyrssen & Wedborg (1974).

to be those exhibiting concentrations greater than $1 \, mg \, kg^{-1}$, but excluding silicon, whose atypical behaviour is influenced by biological processes. Dittmar's analysis of samples collected worldwide during the *Challenger* expedition of 1872–6 revealed that the ratios of the concentrations of the major constituents were invariant with time and location. Such behaviour is termed *conservative* as absolute concentrations are determined by physical rather than chemical or biological processes. Some variations to this observation were noted; in particular Ca^{2+} and HCO_3^- behave *non*-conservatively in deep waters, exhibiting enhanced concentrations due to the dissolution of calcium carbonate. *Salinity*, the total salt content, can crudely be defined as the mass of dissolved solids in sea water expressed in $g \, kg^{-1}$ (‰). The exact definition has evolved historically due to changes in the methodology of salinity measurements and atomic weight scales. Salinity measurements based on the gravimetric analysis of sea salt were fraught with difficulties due to the oxidation of organic matter, volatile loss of CO_2 and halides, and the retention of water of crystallisation. However, the concept of the relative constancy of composition of sea water enabled salinity to be expressed in terms of an easily measurable parameter. This originally was taken to be the *chlorinity* (*Cl*‰), determined by a silver nitrate titration which cannot differentiate between the halides present. As a result of this technique, the definition of chlorinity and hence salinity assumes all bromides and iodides are replaced by an equivalent quantity of chlorides. This is manifest in the ratio of chloride concentration to chlorinity (Table 5.3), that is, $Cl^- : Cl$‰ < 1. A relationship for salinity and chlorinity was derived empirically as:

$$S‰ = 1.805 Cl‰ + 0.030$$

This relationship was determined in 1902 on the basis of gravimetric analysis and argentometric titration of only nine sea water samples, most of which were from the Baltic or Red Sea and hence are atypical oceanic waters. Later, the relationship was redefined using a large number of more representative oceanic waters:

$$S‰ = 1.80655 Cl‰$$

The salinity is now defined in terms of the conductivity of an accurately prepared standard solution of KCl, the most common technique employed at present, and refractive index.

The relative constancy of composition and apparently constant pH lead to the assumption that the oceans may be treated as a steady-state system. This implies that, for a constituent of interest, the rate of supply balances the rate of removal. Accordingly, a *residence time* can be defined as

$$\tau = \frac{A}{dA/dt}$$

where A equals the total dissolved concentration and dA/dt is either the influx or efflux of element A. This concept may be extended to include dissolved gases, nutrients and the minor elements. The residence time provides a crude measure of the relative reactivity of the elements in sea water (Table 5.4). The elements which form insoluble hydroxides have short residence times (aluminium, iron) while the cationic (Na^+, K^+) species and anionic species (Cl^-, Br^-, $UO_2(CO_3)_2^{4-}$) have longer residence times. Such a treatment ignores rapid, biologically induced recycling processes and the spatial variations (both vertical and horizontal) in the concentration of an element. Rather than using oceanic residence times, residence times may be defined for discrete marine environments such as surface waters or anoxic basins.

5.3.2 Equilibrium processes

Equilibrium processes play a major role in controlling the composition of sea water. Such mechanisms determine the dissolved concentrations of atmospheric gases (and consequently the redox potential and pH), the speciation of major and minor elements, and the solubility of several constituents. While the authigenic precipitation of solid phases from sea water will be considered in Section 5.5, other equilibria will be outlined here.

The solubilities of gases in sea water are determined by Henry's law (Section 1.4.7). As a result of the exchange of gases across the air–sea interface, the surface layer of the ocean achieves, or approaches, saturation. Subsequent mixing with deeper waters distributes these atmospheric gases throughout the ocean. Some gases behave conservatively (argon, xenon, N_2 for the most part) with concentrations close to saturation throughout the ocean. Others, notably O_2 and CO_2, exhibit non-conservative behaviour as concentrations are affected by biological processes. These two gases determine the redox potential and pH, respectively, in sea water. The redox potential in oxygenated waters is often taken to be controlled by the couple:

$$O_2 + 4H^+ + 4e^- \rightleftharpoons 2H_2O$$

At 20°C ($K = 10^{83.1}$) and pH 8.1, water in equilibrium with the partial pressure of atmospheric O_2 ($pO_2 = 0.21$ atm) has $pE = 12.5$.

Sea water exhibits a relatively constant pH in the range 7.8–8.2. The buffering is achieved by the equilibration of a dilute bicarbonate solution (alkalinity ~2.4 meq dm^{-3}) (see Section 2.4.4) with atmospheric carbon dioxide ($pCO_2 = 10^{-3.5}$ atm), with a small contribution to the buffering capacity provided by boric acid. The major equilibria

Table 5.4. *The residence times of some elements in sea water*

Element	Principal species	Concentration (mol dm^{-3})	Residence time (y)
Li	Li^+	2.6×10^{-5}	2.3×10^6
B	$B(OH)_3, B(OH)_4^-$	4.1×10^{-4}	1.3×10^7
F	F^-, MgF^+	6.8×10^{-5}	5.2×10^5
Na	Na^+	4.68×10^{-1}	6.8×10^7
Mg	Mg^{2+}	5.32×10^{-2}	1.2×10^7
Al	$Al(OH)_4^-$	7.4×10^{-8}	1.0×10^2
Si	$Si(OH)_4$	7.1×10^{-5}	1.8×10^4
P	HPO_4^{2-}, PO_4^{3-}	2×10^{-6}	1.8×10^5
Cl	Cl^-	5.46×10^{-1}	1×10^8
K	K^+	1.02×10^{-2}	7×10^6
Ca	Ca^{2+}	1.02×10^{-2}	1×10^6
Sc	$Sc(OH)_4$	1.3×10^{-11}	4×10^4
Ti	$Ti(OH)_4$	2×10^{-8}	1.3×10^4
V	$H_2VO_4^-, HVO_4^{2-}$	5×10^{-8}	8×10^4
Cr	$Cr(OH)_3, CrO_4^{2-}$	5.7×10^{-9}	6×10^3
Mn	$Mn^{2+}, MnCl^+$	3.6×10^{-9}	1×10^4
Fe	$Fe(OH)_2^+, Fe(OH)_4^-$	3.5×10^{-8}	2×10^2
Co	Co^{2+}	8×10^{-10}	3×10^4
Ni	Ni^{2+}	2.8×10^{-8}	9×10^4
Cu	$CuCO_3, CuOH^+$	8×10^{-9}	2×10^4
Zn	$ZnOH^+, Zn^{2+}, ZnCO_3$	7.6×10^{-8}	2×10^4
Br	Br^-	8.4×10^{-4}	1×10^8
Sr	Sr^{2+}	9.1×10^{-5}	4×10^6
Ba	Ba^{2+}	1.5×10^{-7}	4×10^4
La	$La(OH)_3$	2×10^{-11}	6×10^2
Hg	$HgCl_4^{2-}, HgCl_2$	1.5×10^{-10}	8×10^4
Pb	$PbCO_3, PbOH^-$	2×10^{-10}	4×10^2
Th	$Th(OH)_4$	4×10^{-11}	2×10^2
U	$UO_2(CO_3)_2^{4-}$	1.4×10^{-8}	3×10^6

Based on Brewer (1975).

involved are:

$$CO_2(g) \rightleftharpoons CO_2(aq)$$

$$CO_2(aq) + H_2O \rightleftharpoons H_2CO_3$$
$$H_2CO_3 \rightleftharpoons H^+ + HCO_3^-$$
(5.52)

$$HCO_3^- \rightleftharpoons H^+ + CO_3^{2-}$$

At pH 8, the most important species is HCO_3^- with relatively insignificant amounts of CO_3^{2-} and dissolved CO_2 (Figure 5.8). From Le Chatelier's principle it can be appreciated that short term fluctuations may be imposed on the pH due to biological influences on the pCO_2. The removal of CO_2 from surface waters during photosynthesis causes the pH to rise. Conversely, the oxidative degradation of organic matter at depth decreases the pH. A long term upward trend in the atmospheric CO_2 content would have different consequences. Equilibration of the ocean as a whole would be a slow process due to the slow mixing rate between the surface layer (~75 m) and deeper waters. Eventual equilibrium would lead to a decrease in pH (equation 5.52) and an increase in the alkalinity of sea water via the reaction

$$CaCO_3(s) + CO_2 + H_2O \rightleftharpoons Ca^{2+} + 2HCO_3^-$$

Buffering of sea water on a geologic time scale may be provided by the reverse weathering of aluminosilicate minerals. This involves reactions of the type:

Figure 5.8. A species distribution diagram for carbonic acid in sea water of salinity 35‰ at 20 °C.

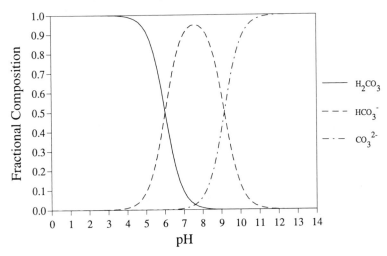

$$3Al_2Si_2O_5(OH)_4 + 4SiO_2 + 2K^+ + 2Ca^{2+} + 9H_2O \rightarrow$$
$$2KCaAl_3Si_5O_{16}(H_2O)_6 + 6H^+$$

Such reactions are unlikely to proceed at rates sufficiently fast as to occur in the water column but may be significant during diagenesis.

By assuming that the ocean as a chemical system has attained equilibrium (a good approximation for the major elements as indicated previously), the speciation of the various constituents can be calculated. Sea water is a $3\frac{1}{2}\%$ salt solution comprising mainly NaCl but also containing many other salts. Hence, several ligands are potentially available for ion pair and complex formation. The major metal ions exist in sea water as free ions with a small percentage complexed with sulphate (Table 5.5). The speciation for other trace metals present in sea water becomes far more complicated (Figure 5.9). Organic complexation is particularly important for copper and zinc. Often the free ion activity, the probable bioavailable portion of an element in solution, may comprise only a small percentage of the total activity. This can have important biological consequences. Several trace metals, particularly copper and zinc, are essential to organisms but toxic at high concentration. These limits define a concentration window essential for the survival of an organism. Some marine organisms may be able to exude organic chelators which may detoxify their environments or alternatively sequester metals making them available for transport across the cell membrane, thereby modifying the concentration window to which they are restricted.

5.3.3 Biological influences

The presence of organisms in sea water can upset established chemical equilibrium and modify the concentration of some constitu-

Table 5.5. *The percentage speciation of the major metal ions in sea water*

Element	$[M^{n+}]$	$[MSO_4]$	$[MHCO_3]$	$[MCO_3]$	$[MB(OH)_4]$	$[MF]$
Na	97.6	2.4	<0.1	<0.1	<0.1	<0.1
Mg	89.0–92.0*	7.8–10.8*	0.1	0.1	<0.03	0.063
Ca	89.0–92.0*	7.8–10.8*	0.1	0.1	<0.03	0.014
K	98.8	1.2	<0.1	<0.1	<0.1	<0.1

* Dependent upon value used for K. $K MgSO_4 = 10.2$ or 7; $K CaSO_4 = 10.8$ or 7.
From Dyrssen & Wedborg (1974).

ents. Such effects have been outlined in terms of pH and the free ion activity of some trace metals but can be extended to include dissolved gases (O_2, CO_2 and occasionally N_2) and the micronutrients (nitrogen, phosphorus, silicon). Consider the reaction for carbon fixation which can be given as the Redfield equation:

$$106CO_2 + 122H_2O + 16NO_3^- + PO_4^{3-} + 19H^+ \rightarrow$$
$$(CH_2O)_{106}(NH_3)_{16}H_3PO_4 + 138O_2$$

where $(CH_2O)_{106}(NH_3)_{16}H_3PO_4$ represents the average composition of marine phytoplankton. Photosynthesis will be light-limited to the surface waters. Photosynthesis can cause depletion of the nutrients, supersaturation of dissolved oxygen and an increase in pH. Organic debris produced in the surface waters falls through the water column and is oxidatively decomposed (i.e. the reverse reaction above). This regenerates the nutrients, decreases the pH and depletes the dissolved oxygen content. These features are most prominent at a depth of approximately 1 km where profiles of dissolved oxygen reveal a minimum concentration. This dissolved oxygen minimum corresponds to maximum concentrations of dissolved nitrogenous and phosphorous nutrients. Recent refinements of analytical techniques have established that

Figure 5.9. The calculated distribution of Pb(II) in sea water at 25°C and 1 atm. (From Bilinski & Stumm, 1973.)

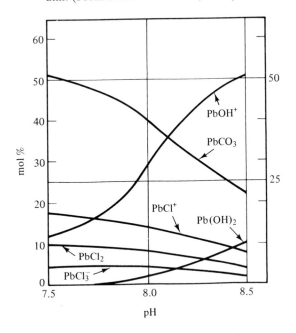

several trace metals also exhibit maximal concentrations at the dissolved oxygen minimum, Figure 5.10. The linear correlations between nickel and phosphate as well as zinc and silicate (Figure 5.11) indicate that biological activity controls their distributions. Under extremely biologically productive surface waters (i.e. eastern tropical Pacific) there is a large flux of organic material to the deeper waters. Subsequent oxidative decay may result in very low levels of dissolved oxygen. In this *oxygen-deficient environment*, mildly reducing conditions may be established, leading to the reduction of nitrate to nitrite. Similar effects may be produced in the localised vicinity of a sewage outfall. Where stagnation of the water column can occur, the oxygen may be com-

Figure 5.10. Dissolved water manganese concentrations observed off central California (36°50′N, 135°00′W) in: December, 1976 (○); April, 1977 (●); July, 1978 (■); and December, 1978 (□). The surface manganese data are those observed while the particle traps were set in December, 1978; other surface values from previous cruises are not shown. Dissolved oxygen data are also from the December, 1978, cruise. (From Martin & Knauer, 1980.)

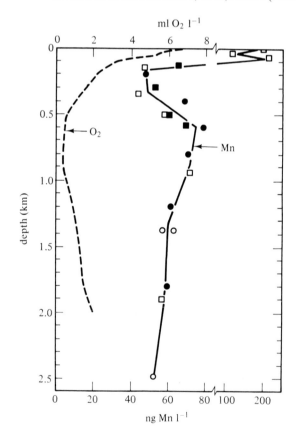

Figure 5.11. The correlation between (a) cadmium and phosphate and (b) zinc and silicate for various stations in the north Pacific. (From Bruland, 1980.)

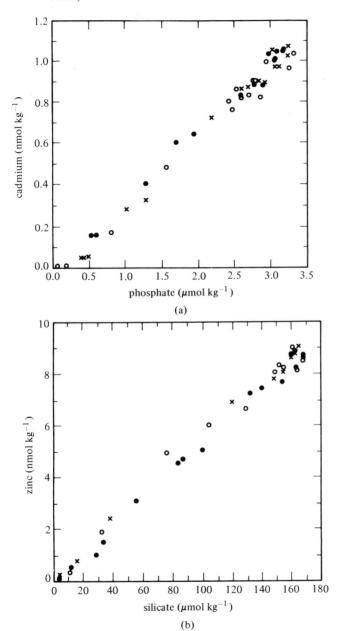

(a)

(b)

pletely utilised producing an *anoxic* environment characterised by strongly reducing conditions (see the following section).

5.3.4 Atypical marine environments

There are a number of atypical marine environments for which the concept of relative constancy of composition does not apply. The concentration of several components (both major and minor elements) may be significantly different to that observed in the bulk sea water solution. These atypical marine environments are generally associated with the boundary regions of the ocean such as the surface microlayer, estuaries, anoxic basins and hydrothermal vents.

While a definitive compositional analysis of the sea surface microlayer remains an elusive goal, the chemical composition of this veneer differs significantly from bulk sea water. Both natural and anthropogenic organic material is concentrated in the surface film, due to its surfactant properties. Enrichment of several trace metals occur, presumably due to the enhanced levels of organic chelators and particulate materials in the microlayer.

From a chemical point of view, an *estuary* is the mixing zone of river and sea waters characterised by gradients in ionic strength and composition. The major influx of most elements to the ocean is via rivers. Material is transported both in dissolved and particulate forms. Estuaries are a complex environment chemically because interactions between these phases may occur, consequently modifying the nature and fluxes of riverborne material to the sea. River water can be generalised as a slightly acidic, low ionic strength solution with a salt matrix of $Ca(HCO_3)_2$ (approximately $120\ mg\ kg^{-1}$). In contrast, sea water is a $35\ g\ kg^{-1}$ Na Cl solution (ionic strength 0.7) with a pH in the range 7.8 to 8.2. Consequently, the salt matrix is determined through most of the mixing zone exclusively by the sea water end member and salinity can be used as a conservative index. A dissolved constituent which exhibits conservative behaviour can be identified as its concentration will vary linearly as a function of the salinity, with a negative or positive slope depending upon the relative concentrations in the river and sea water end members. Positive deviations above the theoretical dilution curve indicate an input of the constituent, while negative deviations indicate some removal process. The major cations behave conservatively while non-conservative behaviour (negative deviation) is exhibited by iron and aluminium due to the precipitation of colloidal hydroxide species.

Manganese is released from estuarine sediments (suspended and bed-load), giving rise to maximum concentrations in the low salinity zone. The pH distribution in an estuary is characterised by a minimum in the initial mixing zone due to changes in the first and second apparent dissociation constants of carbonic acid as a function of salinity. Specia-tion changes occur to dissolved metals due to the increase in type and concentration of available ligands. Chlorocomplexes of cadmium, mer-cury and zinc become significantly more important as the salinity increases. Humic complexation of manganese and zinc decreases due to the mass action effect of a high concentration of calcium and mag-nesium. Finally, flocculation and coagulation processes may remove material from solution. This mechanism may be simply summarised as: particulate material with negatively charged surface may adsorb cations leading to a neutralisation of the surface. As the ionic strength in-creases, there is less electrostatic repulsion when particles collide. Particles aggregate due to weak bonding; this is enhanced with surface coating of organic material.

Atypical marine conditions also exist in *anoxic environments*. When water circulation is restricted both horizontally (due to topographic barriers) and vertically (resulting from thermal or saline stratification), the dissolved oxygen may be exhausted. Typical examples are the Black Sea, permanently anoxic below 200 m, and the Cariaco Trench, a depression in the continental shelf north of Venezuela. Some fjords may be intermittently anoxic (Saanich Inlet in western Canada and Drams-fjord, Norway). Periodic flushing of the inlet with dense, oxygenated waters may displace deep anoxic water to the surface, causing massive fish mortality.

Oxygen is consumed during respiration, but when exhausted, other oxidants are utilised in the following succession:

$$NO_3^- + 2H^+ + 2e^- \rightleftharpoons NO_2^- + H_2O$$

$$NO_2^- + 8H^+ + 6e^- \rightleftharpoons NH_4^+ + 2H_2O$$

$$SO_4^{2-} + 8H^+ + 8e^- \rightleftharpoons S^{2-} + 4H_2O$$

In contrast to fresh waters, marine waters have a relatively high sulphate concentration. Consequently, anoxic conditions develop a sulphide-rich solution which is extremely toxic, and hence these waters are devoid of marine life other than sulphate-reducing bacteria. Secondly, the redox potential is no longer determined by the O_2/H_2O couple but rather by the following equilibrium:

$$HS_2^- + 2e^- \rightleftharpoons HS^- + S^{2-}$$

As a result of the lower redox potential, metals with variable oxidation states will exist in the lower oxidation state, i.e. Fe(II), U(IV) instead of Fe(III), U(VI). This can affect the solubility of these metals. Sediments underlying anoxic waters will be enriched in those metals which form insoluble sulphides (copper, lead) when compared to sediments deposited under oxygenated conditions. However, anoxic waters will be enriched in iron and manganese, metals which form insoluble oxyhydroxides in normal oxic sea water. Table 5.6 summarises those components that are enriched in anoxic waters and related sediments relative to typical sea water.

Sea water can circulate through hot basalt associated with seafloor spreading centres and be discharged as a *hydrothermal spring*. Such features have been observed in the Red Sea, East Pacific Rise and Galapagos Spreading Centre where the hydrothermal water can have a very high temperature, up to 350°C. Red Sea brine pools also exhibit high temperature and salinity (56°C, 257‰). The associated flux of manganese, magnesium and sulphur may be comparable to the riverine supply of these elements. The calculated influx of lithium and rubidium exceeds riverborne fluxes by five and ten times, respectively. The hydrothermal waters have a very high content of sulphide which is quickly oxidised in sea water to sulphate. This supports a dense bacterial population with a carbon fixation rate far in excess of surrounding oceanic waters. This microbial activity serves as the primary

Table 5.6. *Enrichment of constituents in anoxic waters and underlying sediments relative to oxygenated sea water*

Anoxic waters	Sediments
Fe	Cu
Mn	Pb
Phosphate	Zn
Silicates	Ag
NH_4^+	Mo
N_2	U
CO_2	Organic debris
S^{2-}	
Increased alkalinity	
More acidic pH	

trophic level for the unique animal population associated with the hydrothermal vents.

5.4   Soils

5.4.1   Introduction

General composition

Soil is a complex system of solid, liquid and gaseous phases, physically and chemically associated with each other. The solid phase is by far the most heterogeneous component, even in small volumes of soil, being composed of different sized inorganic particles of silica, silicate clay, metal oxides, and other minor components all associated to some extent with organic material.

Soil is formed through the modification of geological material over long periods of time. The interaction of biological, topological and climatic influences in relation to the parent material effect the modification. Igneous rocks, for example granite, formed through the solidification of molten magma and principally composed of a number of alumino-silicates, metal silicates and free silica, may be considered as the primary material of soil formation. Weathering leads to the physical and chemical breakdown and redistribution of igneous rock. Sedimentary rocks formed from the deposition of weathering products of other rocks, for example recemented clays, are termed shales. The influence of heat and/or pressure on igneous or sedimentary material gives rise to metamorphic rocks, for example the hydrothermal alteration of olivine (magnesium iron silicate) forms serpentine.

The comparatively easily weathered feldspars and micas are most common in igneous rocks, and it is their breakdown products which form the fine-grained mineral matter of most soils. The coarse-grained material in soils usually originates from free silica.

Soil description

Soils can be conveniently described by the sequence and nature of the layers (horizons) which constitute the overall structure. This sequence is termed a profile. Horizons within a profile can usually be distinguished visually and have different physical and chemical properties depending on the soil forming process which gives rise to them. There are basically five main groups of horizons distinguishable in soils and these are denoted, O, A, B, C and R horizons and are illustrated in the profile of a podsol (Figure 5.12).

The O horizons form above the mineral matrix and are composed of fresh or part-decomposed organic matter. The mineral A horizons lie at or near the surface and are recognised as zones of maximum leaching. The B horizons are areas below, and often derived from, the A layers where maximum accumulation of iron and aluminium oxides and silicate clays occurs. Enrichment can occur by processes other than downward movement, for example alteration of the original rock material. The C horizons are mineral layers beneath the B horizon. These layers lack any of the characteristics of the horizons above them and may or may not be derived from the material from which A and B horizons were formed. C layers are less weathered and are usually formed under different conditions than the horizons above them.

Figure 5.12. Representation of a podsol profile. The boundaries between O and A and A and B horizons are quite distinct, and good examples of podsols lack A_1 and A_3 zones which, in other soils, represent transitional areas between O and B horizons, respectively.

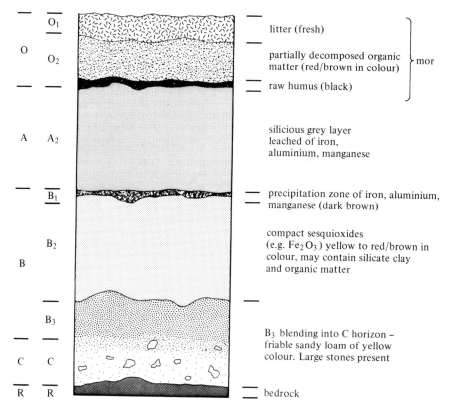

Finally, the R horizon represents bedrock from which all the other horizons may or may not have been derived (i.e. soils may develop from drift material overlying the bedrock).

Specific soil types

When attempting to label a soil as a particular type it must be remembered that we are most likely looking at a material which lies within a continuum of properties and characteristics and will not fit nicely into any rigid set of parameters used to describe a 'classical' soil type. The description of certain soils which follows serves only to illustrate extremes in a wide spectrum of types.

Podsols and brown earths
A generalised profile of a podsol is shown in Figure 5.12. This soil usually develops in well drained locations in areas of high rainfall and cold climate. The profile has a thick dark humus layer (mor) on its surface often subdivided into regions of varying stages of decomposition. There is a sharp division between the mor and the next horizon consisting of pale coloured mineral soil from which most of the iron, aluminium and many other elements have been leached. Immediately beneath the leached A horizon is a very dark B horizon consisting of reprecipitated components from horizons above. The colour of this alluvial horizon gradually lightens towards the C horizon. Each of the main horizons is subdivided into layers of variable depth composed of material characteristic of that horizon but having slight modifications depending on the conditions prevailing during formation of the soil. These have been designated A_1, A_2, etc., in Figure 5.12. The depth of each horizon is very variable and the total depth of the profile (from top of A_0 to bottom of B) may range from a few centimetres to well over a metre. Podsols often form beneath coniferous forests. Often, the alluvial B horizon consists of layers of iron oxides (the 'iron pan') which can prevent root penetration.

There are many forms of podsol found throughout the world, the nature of each depending on climatic conditions, the extent of leaching of the A horizon, and the composition of the parent material from which the soil originated, as well as biotic factors such as the type of vegetation overlying a rich mor. Polyphenols derived from living vegetation (wash off from tree leaves) and mor can disperse or deflocculate clay minerals, causing them to move down the A horizon and accumulate in the B horizon. The slaking or dispersal of clay particles in percolating water when soil is first wet after a dry period is another important podsol forming process. If the soil is low in clay minerals (as in quartz sands) then the polyphenols cause considerable leaching of iron and aluminium, thereby promoting formation of a typical podsol. However, with higher clay content, much of the ferrous polyphenols are adsorbed by the clay, thereby preventing extensive leaching. The soils so formed are not true podsols but are said to be 'podsolised', and are termed brown earths or brown forest soils. The B horizons of these soils have stable crumbly structures due to the coating of iron and aluminium hydroxides on the clay particles. Whereas podsols have a clearly defined organic horizon (mor), which is not mixed with any mineral material, brown earths have the bulk of the humus mixed with mineral soil with only a distinct thin layer of recently deposited material. This constitutes the mull of brown earths and results largely because of an active earthworm population which mixes various horizons together. The acidity of mor in podsols is too extreme for earthworms to be encouraged. Earthworms in brown earths, and all soils for that matter, counteract leaching to a marked extent by depositing soil, ingested from deep horizons, upon the soil surface.

Grassland soils

Grasslands occur on many soil types, but there is one group developed on medium-textured, often calcareous, superficial deposits, formed during the Pleistocene era, which are distributed throughout North America, central and eastern Europe and central Asia which are of particular importance. These soils have been divided into prairie, chernozem and chestnut soils.

Prairie soils are similar to brown earths in many respects. Base cations can be leached from the upper horizons causing them to be slightly acidic, clay and organic matter can be washed down the profile and clay can be formed in deep horizons due to chemical weathering. Mineral pans can be formed as in podsols. These are fertile soils and have been put to good agricultural use.

Chernozems, or black earths (so called because of a black layer of mixed organic and mineral material which may exceed one metre in depth), occur in semi-arid regions where sufficient rain falls to cause some leaching of calcium from surface horizons making them slightly acid. Some of the calcium is precipitated in lower horizons, some two metres down, as calcium carbonate occurring as thin threads or films, and occasionally beneath these threads calcium sulphate is deposited as gypsum. Chernozems represent the principal agricultural soil for wheat production in the world.

Chestnut soils occur in arid regions where the transpiration rate exceeds the rainfall so that water never leaches out of the soil profile. Calcium carbonates and sulphates are precipitated at shallower depths and the arid conditions result in a reduced grass yield.

Changes in land use can modify soil composition. Grasses, herbs and deciduous forest take up large amounts of basic cations during the growing season and release them again to the soil in autumn during leaf fall, thereby making them available for uptake in the following growing season. This cycling of elements reduces nutrient loss through leaching. Growth of coniferous forest tends to encourage leaching due to the acid litter these trees deposit and the removal of nutrients from the foliage prior to leaf abscission rendering the litter low in basic cations. Growth of deciduous forest on podsols can reverse podsolisation.

5.4.2 Composition of the mineral matrix

The major structure of colloidal clay particles is that of layers of plates or flakes. The individual size and shape of the laminations is largely determined by the developmental conditions and the type of mineral concerned. This plate-like structure and finely divided state gives clays a very large specific surface area; for example, the external surface area of 1 g of colloidal clay is approximately 1000 times that of 1 g of coarse sand. As we will see later, this large surface area is of great importance for the adsorption and exchange of ions.

Initially, clays were considered to be amorphous, but X-ray and electron microscope examination has indicated their crystalline nature. Clay minerals occurring in soils are constructed out of sheets containing oxygen or hydroxyl ions and are often termed 'layer-lattice' minerals. A sheet is composed of a group of ions lying in a plane and a layer is built up from a number of sheets having a definite orientation to each other.

The oxygen-containing sheets occur in only two forms. The first is termed a 'complete' sheet since it comprises regular rows and columns of oxygen atoms, each atom having six neighbours in the same sheet. The second, often termed a 'hexagonal' sheet, is similar to a 'complete' structure except that one quarter of the oxygen atoms is missing and each oxygen only touches four others in the same sheet. Hydroxyl ions often replace oxygens in a complete sheet but not in hexagonal sheets.

If a complete (Figure 5.13) and hexagonal (Figure 5.14) sheet are placed together (Figure 5.15), three oxygens of the former will touch one oxygen of the latter, giving rise to a tetrahedral hole between the four atoms (Figure 5.16). In layers built up from these two sorts of sheet, all tetrahedral holes contain a cation (usually silicon, but occasionally aluminium) and are said to be in four-coordination with oxygen. When two complete sheets are stacked together (Figure 5.17) each oxygen in the first sheet is in contact with three oxygens in the second sheet. Thus an octahedron is formed between six oxygens, three oxygens in each sheet (Figure 5.18). Therefore, in addition to tetrahedral holes, this type of layer has octahedral holes. In clays formed from the stacking of complete sheets, none of the tetrahedral holes contain a cation whereas between two thirds to nearly all the octahedral spaces may be filled, usually with aluminium or magnesium.

There are a large number of clay minerals occurring in soils and only a few named examples are given in the discussion that follows. All clay minerals are grouped into two categories – the 1:1 or 2:1 groups

Figure 5.13. Representation of a clay complete sheet.

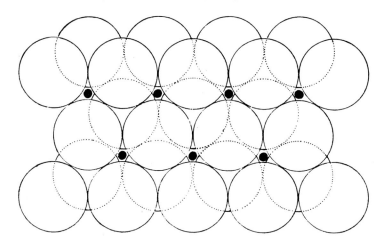

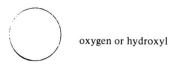

oxygen or hydroxyl

cation

depending on the layer structure. The 1:1 group are built from three sheets – one hexagonal and two complete sheets (the latter one considered as one unit hence the 1:1 term). The 2:1 group have symmetrical structures of two complete sheets sandwiched between hexagonal sheets. The extent to which oxygen is replaced by hydroxyl in the complete sheets within each group varies.

Kaolinite is the archetypical mineral of the 1:1 group of clays. The hexagonal and two complete sheets (considered as a single unit) of each crystal unit are held together by the sharing of oxygen atoms between the silicon and aluminium in the respective sheets. Individual crystal

Figure 5.14. Hexagonal sheet.

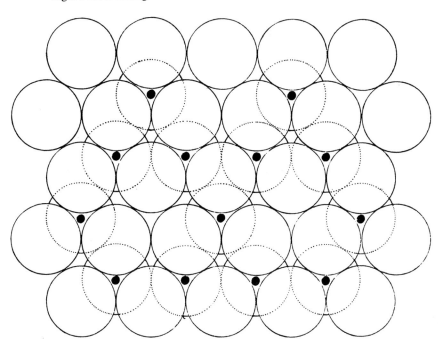

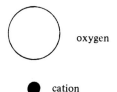

oxygen

cation

units are stacked together to form plates, each crystal unit being held to its neighbour via hydrogen bonding. The hydrogen bonding originates through the interaction of hydroxyl hydrogen in the sheet of one crystal unit with oxygen in a sheet of a neighbouring unit. This results in a rather rigid overall lattice structure which prevents water and cations

Figure 5.15. Stacking of a complete and hexagonal sheet.

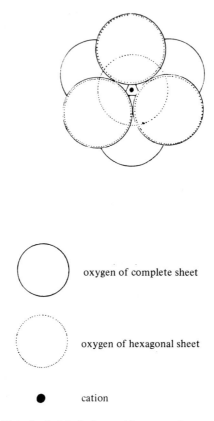

○ oxygen of complete sheet

⋯ oxygen of hexagonal sheet

● cation

Figure 5.16. Tetrahedral hole formed between three atoms in the complete sheet and one atom in the hexagonal sheet.

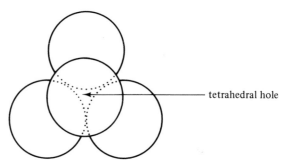

tetrahedral hole

from entering between the structural units. This, coupled with a small surface negative charge, is one of the reasons for the low (compared with other silicate clays) adsorption capacity of kaolinite. Kaolinitic clays can have specific surface areas in the range 5–100 $m^2 g^{-1}$. In general, kaolinite does not exhibit colloidal properties to any great extent.

2:1 minerals have higher negative charges associated with the crystal units than 1:1 structures. Three classes of 2:1 minerals are distinguished, the micaceous clays, vermiculites and montmorillonites, according to the decreasing charge they carry. The crystal units of 2:1 clays are held one to another by electrostatic interactions between surface negative charges in the outer sheets of one unit and the positive

Figure 5.17. Stacking of two complete sheets. Solid spheres are components of the first sheet, and dotted spheres of the second sheet.

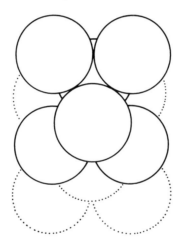

Figure 5.18. Octahedral hole formed between three atoms of each complete sheet.

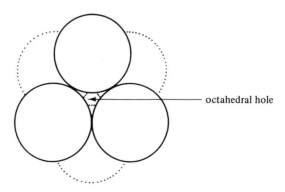

charges in sheets of other crystal units. In micaceous and vermiculite clays the force of attraction between crystal units can be strong. Layers in montmorillonite, having the smallest surface negative charge, are only weakly attracted to each other, making it easy for the mineral structure to expand allowing water and cations to move between the crystal units. Thus, the area exposed for cation exchange is greatly increased. Montmorillonite has about ten to fifteen times the cation adsorption capacity of kaolinite. If all the surface negative charges on montmorillonites are neutralised by cation adsorption, the attractive forces between crystal units are so weak that water can penetrate between individual layers and the clay structure swells. A well dispersed sodium montmorillonite can have a specific surface area in the range 700–800 $m^2 g^{-1}$.

In the micaceous clays, the greater negative surface charge (due to the replacement of about 15% of the silicon by aluminium) is neutralised by potassium ions which fit into the hexagonal holes in the crystal lattice surface. This imparts stability to the crystal lattice and results in a less expansive structure than, for example, montmorillonites. Micaceous clays typically have specific surface areas in the range 100 to 200 $m^2 g^{-1}$. Micaceous clays are sometimes referred to as illites, although this is not strictly correct since true illites consist of a mixture of clay minerals and are not pure micaceous clays.

There are clays in which most of the silicon has been replaced through weathering processes by aluminium and iron. These may occur in temperate regions although iron and aluminium oxides are particularly important in tropical and semi-tropical regions.

This section has described the mineral matrix of soils and has outlined basic composition of major clay minerals. It should be remembered that 'pure' clay minerals probably only exist in unweathered rock material and that in soils different sorts of clays will often be mixed together. In addition the surfaces of clay particles are often covered with films of hydrated oxides of aluminium and iron. Another very important material intimately associated with clay materials in soils is organic matter.

5.4.3 Organic matter

Soil organic matter consists of a number of materials ranging from undecomposed plant or animal tissue through various intermediate levels of decomposition to a comparatively stable, amorphous brown/black material. The latter is usually considered as soil humus. This is often split into three categories: (a) fulvic acid (low molecular weight material soluble in both acid and alkali); (b) humic acid (intermediate molecular weight material, soluble in alkali but not acid); (c) humin (highest molecular weight fraction and insoluble in both acid and alkali). The colour of each fraction darkens from (a) to (c). Much or most of the humic material is bound to clays through either the sharing of polyvalent cations, for example calcium or aluminium, through interaction of the negatively charged humus colloids with the positively charged surface of iron or aluminium hydrated oxides or via hydrogen bonding or van der Waals forces. These interactions give rise to the 'clay/humus complex'.

Humic colloids are essentially composed of carbon, hydrogen, oxygen, nitrogen, phosphorus and sulphur, the ratio of each depending on the soil type. In bulk agricultural soils it is common to find a carbon:nitrogen:sulphur:phosphorus ratio of $100:10:1:2$.

A significant component (5–20%) of soil organic matter is composed of carbohydrates. These are recruited to soil in comparatively large amounts from plant residues, 50–70% of plant dry matter being composed of carbohydrate. Other important sources are fungal and bacterial metabolites and their decomposition products. The majority of carbohydrate is in the form of polysaccharides being composed of monomer units of sugars such as glucose, galactose, mannose, arabinose, xylose and a number of others (see Chapter 3). A very small and quantitatively insignificant amount of free monosaccharides occurs in soils, being somewhat more abundant in cool climatic conditions.

Other components of organic matter from plant and animal sources such as proteins, amino acids and lipids are generally broken down quickly by soil microorganisms. Cuticular waxes however may persist for a considerable time and may form fossil deposits.

Degradation products of plants and animals and the synthetic products of microorganisms interact forming complex structures more stable than the source material. The greatest proportion of soil organic matter is in the form of humic complexes – principally of phenolic polymers – the structural units of which resemble those of lignin. In addition complexes can be formed with metal ions and hydrous oxides. For example, fulvic acids can form water soluble complexes with toxic metals, thereby elevating metal concentrations in natural waters and soil solutions.

Soil organic matter is of critical importance in giving soil its structure, i.e. an aggregation of mineral material with pore spaces providing aeration. The clay/humus complex is thought to arise through the binding of clay particles by organic macromolecules as shown below

R–COO–Ca–clay

The amount of organic matter in soil can vary from <1% in entisols (recent soils) to as much as 90% in histosols (bog/peak soils). The degree of mixing of organic and mineral material is also variable. Podsols have a clearly defined organic layer on their surface whilst mollisols (which include important groups of agricultural soils) have a thorough mixing of mineral and organic constituents. In the latter, the contribution of organic matter towards the cation adsorption within the soil is small compared with that due to clay minerals. However, in sandy soils, organic material may be the dominant factor responsible for cation adsorption.

5.4.4 Cation exchange

Cation exchange is a consequence of the negative charges on colloidal clay and humus particles of the soil. The negative charge has two principal components, a permanent charge and a variable pH dependent charge. The permanent charge arises because of isomorphous substitution within mineral lattices where aluminium (Al^{3+}) substitutes for silicon (Si^{4+}) in tetrahedral sheets, and magnesium (Mg^{2+}) or iron (Fe^{2+}) can replace aluminium in octahedral sheets. The solid fraction of the soil as a whole, because of diminution of positive charge, develops a net negative charge.

The variable charge component results through the dissociation of hydroxyl groups present on alumino-silicate surfaces or in phenolic or carboxylic groups. Hydroxyl dissociation is a pH dependent process tending to occur at high pH values (>pH 7) in phenols but can occur at pH 3 to 4 in carboxylic acids associated with organic matter. Where appreciable amounts of organic matter are present, such as in surface soils, the major part of the variable charge arises through proton dissociation from carboxylic acid and phenolic groups. Negative charge therefore, within a particular soil, depends not only on silicate clay minerals but also on pH and organic matter content, both of which may be manipulated by agricultural practice.

With the development of negative charge within a soil, positively charged counter-ions (mainly Ca^{2+} but also K^+, Na^+, NH_4^+, Mg^{2+}, Mn^{2+} and Al^{3+} depending on circumstances) are adsorbed by soil particles, thereby maintaining overall electrical neutrality. There is competition between cations for the negatively charged sites (exchange sites) on the soil matrix such that cations freely available in the soil solution, which bathes the solid matrix, may exchange with cations held on the exchange sites. The distribution of each cation between soil surface and soil solution is controlled by

(a) the charge, size and hydration state of the cation,
(b) the density of negative charge on the soil surface,
(c) the ionic strength of the soil solution,
(d) selective minerals on size criteria or where organic-complexing may occur.

The amount of negative charge on soil solid matrices is quantified in terms of the cation exchange capacity and has units of milliequivalents per unit mass (meq kg^{-1} soil). This is a measure of the ease with which a soil can adsorb cations and desorb them when in the presence of competing ions.

Ion-exchange mechanisms form an important part of our understanding of many environmental and ecological processes. In the following section we will consider a particular process, that of soil acidification, through which the concept of soils as a dynamic ion-exchange system will be expanded.

5.4.5 Soil acidification

In recent years, increased industrial emissions of sulphur dioxide (SO_2) have resulted in adverse effects on terrestrial and aquatic ecosystems. Initially the direct effect of SO_2 and other gaseous pollutants (for example mixtures of nitrogen oxides – often referred to as NO_x) were considered, but now the situation has been compounded by the indirect influence of SO_2 oxidation products being deposited as 'acid rain' in wet deposition and as acidic particles in wet and dry deposition. Acidity in rainfall has been the focus of much research activity in recent years because of the complex and often disastrous way in which ecological systems may be affected.

Acid rain has been implicated in adverse effects upon vegetation (affecting productivity of, for example, forests), soils (accelerating leaching of cations and reducing soil fertility) and aquatic systems (adversely affecting fisheries). Soils and aquatic systems are intimately linked since much of the water recruited to water systems has been in long or short term contact with soils or parent rock material.

The following discussion deals with the acidification of soils, the sources of acidity in the soil environment and how soil leachates can be modified by acidification processes.

Sources of acidity

Acid precipitation

Acids and potentially acidifying substances reaching soils from the atmosphere may be deposited wet (in rainfall) or dry (gaseous plus particulate) and include the following:

> sulphur compounds – SO_2, SO_3^{2-}, SO_4^{2-}, H_2SO_4,
> nitrogen compounds – NO, NO_2, NO_2^-, NO_3^-, HNO_3, NH_4^+,
> chlorine compounds – HCl,
> others – weak acids (organic acids, H_2CO_3), Brønsted acids
> (e.g. dissolved Fe and NH_4^+).

Acidity in rainfall is mainly attributable to the strong mineral acids H_2SO_4 and HNO_3 which are derived from industrially produced SO_2 and NO_x, respectively. In the days before the Clean Air Acts were enacted, the comparatively high dust load (the bulk of which was alkaline in nature) in the atmosphere tended to neutralise and thereby diminish the acidity. Now that this dust loading is much reduced, the atmosphere has a lower neutralising capacity, and acidity is free to reach the terrestrial system, possibly having originated from a source many hundreds of kilometres away. It is not uncommon to have a rainfall of $1000\,\text{mm}\,\text{y}^{-1}$ in Britain, which if acidified to pH 4 (which is quite common) would contribute $100\,\text{meq}\,\text{H}^+\,\text{m}^{-2}\,\text{y}^{-1}$.

Acidity from soil organic matter deposition

Coniferous forests can produce an acid litter since the tissues of such vegetation contain considerable concentrations of soluble organic acids. Fulvic acids in soil humic material can be readily leached by dilute acids, and along with the soluble organic acids from vegetation can produce strong acidification and weathering of soils leading to podsol formation and low base saturation. Oxidation of nitrogen and sulphur compounds associated with organic matter can acidify soils, along with natural and fertiliser applied NH_4^+ which is slowly oxidised to nitric acid. These processes are attributed to soil microflora and fauna. Artificial drainage of soils rich in reduced sulphur compounds, for example hydrogen sulphide (H_2S) in peat bogs, can cause acidification as the sulphur compounds become oxidised when exposed to the atmosphere.

Acidity due to dissolved CO_2
Respiratory CO_2 from plant roots and soil microorganisms in the soil solution will form carbonic acid (H_2CO_3) which will dissociate to give HCO_3^- and H^+. This process is appreciable above pH 5. In calcareous soils, H^+ recruitment from H_2CO_3 can be as high as 1000 meq m^{-2} y^{-1}.

Nutrient uptake by plants
Cations removed from soil solutions by plant roots are exchanged for H^+ ions. This leads to acidification of soils but has proved very difficult to quantify.

Leaching processes

The chemical composition of water draining from the base of a soil profile will usually bear very little resemblance, in either a qualitative or quantitative sense, to that of rain water entering soil from the atmosphere. We have seen that there are a number of sources of hydrogen ions in the atmosphere and soil environment which may recruit acidity to soil solutions. We will now consider how this acidity, along with cations and anions in soil solution, interacts with the soil solid phase, thereby bringing about changes in the composition of both soil and soil water draining from beneath horizons.

Easily soluble minerals
A common example of a soil mineral which can easily be taken into solution through the action of acidic percolation water is found with calcite ($CaCO_3$). Disintegration and decomposition of the mineral is accelerated by the presence of hydrogen ions, even those associated with weak acids such as carbonic acid. This acid is known to result in the chemical solution of calcite in limestone the outline of which is given below (equations 5.53–5.55):

$$CO_2 + H_2O \rightarrow H_2CO_3 \tag{5.53}$$

$$H_2CO_3 \rightarrow H^+ + HCO_3^- \quad (pK = 6.37) \tag{5.54}$$

$$CaCO_3(s) + H_2CO_3 \rightarrow Ca^{2+} + 2HCO_3^- \tag{5.55}$$

where (s) = solid phase. The equilibrium condition for this process is given as

$$\frac{\{Ca^{2+}\}\{HCO_3^-\}^2}{\{CaCO_3(s)\}\{CO_2\}\{H_2O\}} = K_1 \tag{5.56}$$

Taking the activities of solid calcite and soil water to be unity, equation (5.56) will therefore simplify to

$$\frac{\{Ca^{2+}\}\{HCO_3^-\}^2}{\{CO_2\}} = K_1 \qquad (5.57)$$

Calcite solubility is therefore enhanced by increasing CO_2 concentration in soil air spaces and inhibited by increasing Ca^{2+} activity. The atmospheric CO_2 concentration is about 0.03% but in soil may rise to 10%, particularly in surface soils and upper regions of sub-soils due to microbial activity and plant root metabolism. Increasing Ca^{2+} contents of soil may arise through fertiliser application or atmospheric deposition. Acid precipitation is an additive process to calcite (and mineral dissolution in general) dissolution caused by H_2CO_3 produced in soils. The effect of acid precipitation on calcite can be expressed as equation (5.58) and the equilibrium condition by equation (5.59):

$$CaCO_3(s) + H^+ \rightarrow Ca^{2+} + HCO_3^- \qquad (5.58)$$

$$\frac{\{Ca^{2+}\}\{HCO_3^-\}}{\{H^+\}} = K_2 \qquad (5.59)$$

All bicarbonate ion is produced from the calcite. If sulphuric acid is the deposited acid in rainfall then calcium sulphate will be produced (equation 5.60). This would be an additional factor to be considered in the dissolution of less soluble minerals such as gypsum itself ($CaSO_4$):

$$2CaCO_3 + H_2SO_4 \rightleftharpoons Ca^{2+} + 2HCO_3^- + CaSO_4 \qquad (5.60)$$

Sparingly soluble minerals
The dissolution of particularly insoluble minerals is generally referred to as weathering, although the term can also be applied to easily soluble minerals. The stronger mineral acids present in rain water and fulvic and other organic acids derived from organic matter promote weathering as does the acidity of the soil solid phase itself. Hydrogen ions in percolation water can replace potassium in biotite (and other mica-related material) fairly easily resulting in the soil itself becoming acidified and soil water becoming enriched in potassium:

$$\boxed{Mica} - K + H^+ \rightarrow \boxed{Mica} - H + K^+ \qquad (5.61)$$

Other elements (calcium, magnesium, sodium) can be transferred to soil solutions in this way from minerals such as augite and hornblende. If the amount of hydrogen ion in percolating water increases, from whatever source, and remains high, the silicate structure of minerals within soil is gradually destroyed, liberating silicic acid as well as free cations (equation 5.62). Silicic acid has a pK of 9.6 and will remain substantially undissociated at the pH encountered in most soil solutions:

$$2KAlSi_3O_8 + 2H_3O^+ + 7H_2O \rightarrow H_4Al_2Si_2O_9 + 2K^+ + 4H_4SiO_4$$

feldspar   hydroxonium ion   kaolinite   silicic acid

$$(5.62)$$

If the soil becomes very acidic then aluminium associated with clay minerals may become soluble in the form of Al^{3+} or aluminium hydroxy cations. Soluble aluminium may also be recruited to percolation water through organic matter decomposition. These ions may then become adsorbed even in preference to hydrogen ions, present in soil solution, by the permanent (pH independent) negative charge in soil clay mineral surfaces. The adsorbed aluminium is in equilibrium with aluminium ions in the soil solution. Soil solution aluminium ions can contribute to soil acidity through their tendency to hydrolyse. This is illustrated in equations (5.63) and (5.64).

$$
\boxed{\begin{array}{c}\text{clay mineral}\\\text{and organic matter}\\\text{aluminium}\end{array}}
\underset{\text{high pH}}{\overset{\begin{array}{c}\text{low pH}\\(\sim<\text{pH }3)\end{array}}{\rightleftharpoons}}
\quad Al^{3+} \qquad\qquad (5.63)
$$

(solid)   (soil solution)

$$Al^{3+} + H_2O \rightleftharpoons Al(OH)^{2+} + H^+ \qquad (5.64)$$

(soil solution)   (adsorbed to   (increases
                permanent charge   acidity of soil
                sites in preference   solution)
                to H^+)

Thus, under particularly acidic soil conditions, complex equilibrium processes can occur between H^+, Al^{3+} and aluminium hydroxy compounds on exchange sites and in soil solution. Similar reactions occur with iron and manganese present in clay minerals. In very acid soils the concentration of aluminium, iron and manganese in soil water draining from the base of soil profiles can be very high. This leads to the precipitation of gelatinous metal hydroxides ($Al(OH)_3$ and $Fe(OH)_3$) in water courses of high pH.

As the pH of percolating water increases then aluminium in soil solution can no longer exist as free Al^{3+} ions but is converted to hydroxy ions as shown below:

$$Al^{3+} + OH^- \rightleftharpoons Al(OH)^{2+} \tag{5.65}$$

$$Al(OH)^{2+} + OH^- \rightleftharpoons Al(OH)_2{}^+ \tag{5.66}$$

As in very acid soil conditions, the hydroxy compounds can be held on exchange sites. In solution they can undergo further reactions by hydrolysis:

$$Al(OH)^{2+} + H_2O \rightleftharpoons Al(OH)_2{}^+ + H^+ \tag{5.67}$$

$$Al(OH)_2{}^+ + H_2O \rightleftharpoons Al(OH)_3 + H^+ \tag{5.68}$$

Thus, although percolating water entering the soil mineral horizons may have a progressively increasing pH, leachate water leaving the soil will maintain a low pH as reactions such as (5.67) proceed and release hydrogen ions to the soil solution through hydrolysis and the exchange of hydrolysis products ($[Al(OH)_2]^+$) for H^+ on clay mineral exchange sites:

$$\boxed{\text{clay mineral}}^{-\,+}H + Al(OH)_2{}^+ \rightleftharpoons \boxed{\text{clay mineral}}^{-\,+}Al(OH)_2 + H^+$$

   (solid phase)   (solution)   (solid phase)   (solution)

$$\tag{5.69}$$

In some 2:1 type clays, particularly vermiculite, the aluminium hydroxy ions (and iron hydroxy ions) can move between crystal units and become very tightly bound. In doing so they prevent inter-crystal expansion and block some of the exchange sites. They can be removed by raising the soil pH. The exchange sites will then become available for further exchange reactions. In this way, aluminium hydroxy ions can be considered to influence the pH dependent charge component of soil colloids.

If the pH of percolating water reaching the mineral horizons rises (occasionally becoming neutral or alkaline) then pH dependent and independent exchange sites will liberate H^+ into the soil solution in exchange for base cations such as calcium. Thus the leachate water will continue to have a lower pH than water entering the soil but the soil itself will become less acidic and increase in base cation saturation.

Under such circumstances, all the hydroxy aluminium compounds will have been converted to gibbsite:

$$Al(OH)_2^+ + OH^- \rightarrow Al(OH)_3$$
$$\text{(gibbsite)}$$

(5.70)

5.4.6 Ion exchange and acidification

The base cation status of a soil and its soil solution will be determined by a number of interacting factors. Figure 5.19 is a schematic representation of the principal factors involved in modifying the composition of soil solution, exchangeable base status and leachate water. Exchangeable base cations are Mg^{2+}, K^+, Na^+ and Ca^{2+} as the dominant ion. Major acidic ions are H^+ (or more correctly written H_3O^+), Al^{3+}, $Al(OH)^{2+}$ and $Al(OH)_2^+$, with minor contributions from Fe^{3+}, $Fe(OH)^{2+}$, $Fe(OH)_2^+$, Fe^{2+} and Mn^{2+}. As we have seen, lowering the pH of percolation water results in a rise in the ratio of acid cations to base cations held on exchange sites. Chemical weathering of minerals causes an increase in the ratio of aluminium-hydroxy compounds to hydrogen ions. Base cations tend to be removed from exchange sites in a particular order. This is related to the ease of removal of particular cations from exchange sites through differences in their binding energies to the soil solid phase. Typically, monovalent ions Na^+ and K^+ are more easily exchanged and therefore leached from soils than divalent cations such as Mg^{2+} and Ca^{2+}.

Whilst it has been made clear that acidity in percolation water can exchange for soil base cations and be involved in complex reactions with, for example, aluminium, we have not considered the role of base cations entering the mineral soil system in percolation water. Any neutral salt added to a soil will be involved to some extent in an exchange process. The cation of the salt may replace hydrogen or aluminium ions associated with soil exchange sites and cause a pH drop in the soil solution:

$$\boxed{\text{clay}}^{-\ +}H + KCl \rightleftharpoons \boxed{\text{clay}}^{-\ +}K + HCl \tag{5.71}$$

$$\boxed{\text{clay}}^{-\ +}Al(OH)_2 + KCl \rightleftharpoons \boxed{\text{clay}}^{-\ +}K + Al(OH)_2Cl \tag{5.72}$$

$$Al(OH)_2Cl + H_2O \rightarrow Al(OH)_3 + HCl \tag{5.73}$$

Acid produced by equations (5.71) and (5.73) is called exchange acidity, and its magnitude depends on many factors including bonding energy of salt cation, pH and salt content of the soil and the amount of neutral salt added to the soil. Exchange acidity increases with cation valency and falls with the extent to which ions may be hydrated. The picture is

Figure 5.19. Schematic view of sinks and sources of cations in the soil environment.

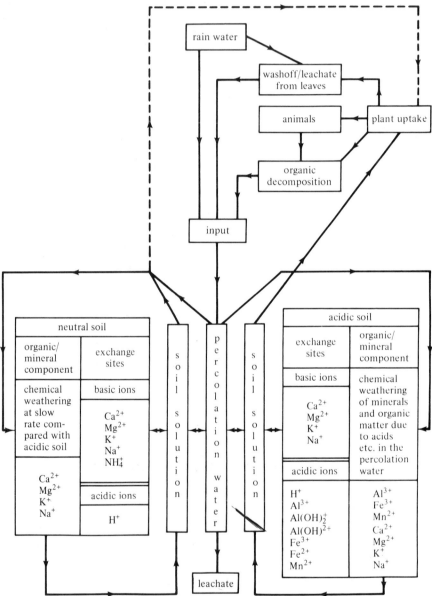

complicated through the amphoteric (i.e. exhibiting both acidic and basic properties) nature of soils. Anions of any salt added to a soil cause exchange with soil basic groups. Insoluble hydrous oxides of aluminium, iron and manganese such as limonite and goethite minerals can undergo exchange reactions as outlined in reaction (5.74) causing a pH increase of soil solutions:

$$(FeOH_2OH)_2 + Na_2SO_4 \rightarrow (FeOH_2)_2SO_4 + 2NaOH \quad (5.74)$$

The base formed is termed exchange alkalinity, and its magnitude depends on similar factors to exchange acidity. The adsorption of sulphate and phosphate probably accounts for most exchange alkalinity.

The effect of neutral salts in soil percolation water on soil solution pH is determined by a balance of exchange acidity and exchange alkalinity. In the majority of situations, the adsorption of anions applied as neutral salts is very small, so that exchange acidity is usually greater than exchange alkalinity.

Thus, application of neutral salts to soils in precipitation results in cations from the salt exchanging for hydrogen or aluminum hydroxides. The base saturation of the soil therefore increases along with its pH, whilst the soil solution leaving the base of the soil profile will become acidified. All basic cations, principally calcium, magnesium, potassium and sodium, irrespective of whether added as fertilisers or deposited from the atmosphere, behave as outlined in equations (5.71) or (5.72). The effects of neutral salts on soil acidification are markedly influenced by the bonding energy of H^+ to the soil exchange sites. At pH > 6 practically all the H^+ in the percolating water is adsorbed by the soil irrespective of the salt concentration supplied. At pH < 6 the influence of neutral salts increases as the pH falls, resulting in reduced soil acidification and a corresponding increase of the leachate acidification.

5.4.7 Percolation water residence time

Water entering a soil system will be in contact with soil organic and mineral material for a finite time. Precipitation entering soil from a sudden downpour may have a rather short contact time with the soil compared with that entering from a light shower. Other factors such as topography and soil depth greatly influence percolation water contact time with soils. Whilst cation-exchange reactions are rapid, occurring within a few seconds in soil suspensions, in naturally occurring soils the exchange rate is much slower due to rate limiting steps such as diffusion rates of ions from soil aggregates. Adsorption of anions such as PO_4^{3-} and SO_4^{2-} probably occurs more slowly than cations so that residence time may be important for their reactions and ultimate exchange within soils.

There are circumstances when precipitation water may pass through a soil into receiving streams and lakes largely unchanged by interaction with soils. This may result through surface runoff when, during heavy downpours, water does not penetrate the soil surface and runs along its surface directly into a receiving water. If soils are very shallow, overlying bedrock, then water entering the soil system may be in contact with very little material with which exchange reactions may occur. Reduced contact time of percolation water and soil due to channelled flow will further reduce the extent to which rain water may interact with the soil. These conditions may be encountered in many parts of the world where unweathered rock is overlain with shallow, base-poor, soils. Rain water, the pH of which may often range between 3 and 4, if not neutralised through exchange processes within the soil, can cause lakes and streams to become acidic, thereby drastically affecting the aquatic ecosystem. Ion-exchange processes are complex, and many of the factors relating to soil acidification are still the subject of intense research activity. The importance of soil in modifying percolation water and the changes that may occur in soil itself through exchange processes has been stressed. Changes in soil through deposition of anthropogenic pollution, or changes in land use and management, not only affect soil fertility but may have a direct or indirect effect on many other components of the environment and have been shown to be linked with the productivity of important parts of the world's ecosystem.

5.5   Marine sediments

5.5.1 Introduction

The ocean floor is covered with a layer of unconsolidated material varying in depth from 0 to 1000 m, with an average thickness of 500 m. Due to plate tectonics, none of the sediments are older than Jurassic (166 million years). Sedimentation rates vary from several millimetres per year in near-shore environments to 0.2–7.5 millimetres per thousand years in the deep-sea. Marine sediments comprise a complex mixture of components derived from different sources and subjected to varying transportational histories. Chemical processes play a major role in determining the nature of these sediments as reactions both in sea water and within the sediments themselves can lead to the formation and/or alteration of sedimentary components. Such chemical processes are considered in greater detail in subsequent sections. However, they cannot be examined in isolation due to the importance of physical, geological and biological interactions. By means of an introduction, a brief but descriptive outline of these non-chemical influences provides a conceptual framework for the subsequent discussion of the chemistry of marine sediments.

These four factors which determine the characteristics of any oceanic sediment are:

  (i) source of the components, by which means the sediments are generally classified;
 (ii) transportation both to and within the ocean;
(iii) deposition;
 (iv) *diagenesis*, that is the post-depositional alteration of the sediments.

Sediments are classified according to the source of the major contributing component. The four classifications of sedimentary components are listed in Table 5.7 together with appropriate examples of mineral phases. It should be noted that several synonyms may be used to define the four categories outlined below.

Lithogenous (terrigenous) material

This is from the continental land masses derived by weathering processes. The relative importance of lithogenous material in the sediments will depend upon their proximity to the continents, especially for riverborne material, and the dilution in the sediments by components from other sources. Considering the time scales involved, lithogenous components can be described as chemically inert in the water column. However, lithogenous material can act as an important substrate for the adsorption of organic matter and trace metals. Similarly, the major cations may be involved in ion-exchange processes, particularly with clay minerals which may undergo reverse weathering reactions in the marine environment.

Hydrogenous (halmeic, chemogenous) material

This is formed in sea water by inorganic reactions. Two processes may give rise, respectively, to primary and secondary material: *authigenic precipitation*, in which new mineral phases are formed directly from sea water, and *halmyrolysis* (formerly known as submarine weathering), whereby sedimentary components of continental or volcanic origin undergo alteration due to low temperature reactions with sea water. Such chemical processes are discussed in greater detail in Section 5.5.2.

Biogenous (biotic) material

This is formed by the fixation of mineral phases by marine organisms. Calcite and silica are quantitatively the most important minerals as they are incorporated into the tests (i.e. skeleton or shell) of several phytoplankton and zooplankton species. Accordingly, the productivity of surface waters together with the species composition of the marine life will play a major part in determining the relative importance

Table 5.7. *The four categories of marine sedimentary components with examples of mineral phases*

Classification	Mineral example	Chemical formula
Lithogenous	Quartz	SiO_2
	Microcline	$KAlSi_3O_8$
	Kaolinite	$Al_4Si_4O_{10}(OH)_8$
	Montmorillonite	$Al_4Si_8O_{20}(OH)_4 \cdot nH_2O$
	Illite	$K_2Al_4(Si, Al)_8O_{20}(OH)_4$
	Chlorite	$(Mg, Fe^{II})_{10}Al_2(Si, Al)_8O_{20}(OH, F)_{16}$
Hydrogenous	Fe–Mn minerals	$FeO(OH)–MnO_2$
	Carbonate fluoroapatite	$Ca_5(PO_4)_{3-x}(CO_3)_xF_{1+x}$
	Barite	$BaSO_4$
	Pyrite	FeS_2
	Aragonite	$CaCO_3$
	Dolomite	$CaMg(CO_3)_2$
Biogenous	Calcite	$CaCO_3$
	Aragonite	$CaCO_3$
	Opaline Silica	$SiO_2 \cdot nH_2O$
	Apatite	$Ca_5(F, Cl)(PO_4)_3$
	Barite	$BaSO_4$
	Organic matter	
Cosmogenous	Cosmic spherules	
	Meteoric dusts	

of biogenous components in the sediments. Biogenous sedimentation is considered in Section 5.5.3.

Cosmogenous (extra-terrestrial) material

This is derived from an extra-terrestrial source. This material comprises small (generally <0.5 mm) black micrometeorites or cosmic spherules. They are composed of magnetite or a silicate matrix incorporating magnetite crystals. Cosmogenous components constitute an insignificant but ubiquitous fraction of the marine sediments. The relative contribution of cosmogenous material increases with a decrease in the sedimentation rate, and for this reason are most abundant in the central Pacific Ocean.

Transportation processes both to and within the ocean will affect the sedimentation rate and composition of marine deposits. Considering firstly the transport of material to the ocean, this process will influence largely the distribution of lithogenous material and to a lesser extent the distribution of halmyrolysates. By far the most important mechanism is river discharge. Of an estimated annual flux of solid material of 2.26×10^{13} kg, about 1.84×10^{13} kg is supplied to the oceans via rivers together with a further 4.2×10^{12} kg of dissolved solids. For any one river, the supply of riverborne material depends upon the size, geology and climate of the catchment area. Most of this lithogenous material will be deposited in the immediate vicinity of the rivermouth, generally situated in a coastal marginal sea. Aeolian or wind transport is quantitatively less significant (annual flux equals approximately 6×10^{10} kg) but is manifest in the zonal distribution of desert dusts, especially the red-brown, iron oxide-coated quartz grains of Sahara Desert origin found in much of the North Atlantic. Aeolian transport also causes the global dispersion of radionuclides produced by weapons testing. Similar to the deposition of cosmogenous material, wind-blown material is relatively most abundant in areas with low sedimentation rates such as topographic highs on the ocean floor and the central Pacific Ocean.

Transportation within the oceans is effected by means of ice-rafting, organisms and currents, including turbidity currents. Both turbidity currents and ice-rafting can introduce relatively coarse sediment to the deep-sea, although ice-rafting is restricted to polar regions. Present limits of ice advance are about 40°N and 55°S, but there have been considerable variations in the geologic record. Organisms are important quantitatively for biogenous sedimentation but also exert a significant influence on fine-grained lithogenous material. Such sedimentary components may become incorporated into faecal pellets, and consequently the settling rates for this fine-grain material are accelerated. Ocean currents are most important for the distribution of particles with long residence times. Main surface currents are zonal and reinforce zonal aeolian supply patterns. Deep-water currents are of little consequence as velocities are low compared to the settling rates.

Deposition naturally follows the cessation of transport of sedimentary components in the ocean. Deposition for biogenous sediments is initiated with the loss of buoyancy of organisms, generally following death. Hydrogenous sedimentation results when the solubility product of a constituent is exceeded. This may be associated with distinct chemical boundaries such as the oxygenated/anoxic interface. Regardless of their origin, the deposition of the sedimentary components is determined by the hydrodynamic features of the environment. That is, deposition occurs when the available energy is insufficient to transport the available load. The settling velocity of individual particles depends upon their size, shape and density.

Finally, the nature of the sediments is influenced by diagenesis, the term applied to refer to the processes which occur within the sedimentary column. These processes may be physical (i.e. compaction associated with the accumulation of thick deposits) or chemical (i.e. authigenic precipitation following compositional changes within the interstitial water). Diagenesis is considered in detail in Section 5.5.4.

5.5.2 Hydrogenous sediments

As outlined previously, hydrogenous sedimentary components may be classified as either primary (precipitates) or secondary (halmyrolysates) material. Authigenic precipitation occurs as the result of a constituent exceeding its solubility product in sea water. This may occur as a result of evaporation, a change in the oxidation state of the constituents and small pH changes for $CaCO_3$ phases, or an excessive supply of material due to volcanism. This classification includes the precipitation of surface coatings onto other sedimentary components, provided such components act only as a substrate for nucleation and themselves undergo no chemical reactions.

Important authigenic precipitates are ferromanganese nodules, metalliferous sediments associated with hydrothermal activity, aragonite [$CaCO_3$], barite [$BaSO_4$] and apatite [$Ca_5(F, Cl)(PO_4)_3$]. Halmyrolysates result from the alteration of sedimentary components due to low temperature reactions with constituents dissolved in sea water. This classification includes the alteration products formed when lithogenous material encounters sea water in estuarine environments or following lava–sea water interactions.

Ferromanganese nodules

Ferromanganese nodules are probably the best known marine mineral. This stems from their considerable potential as an exploitable mineral resource and the controversy of their formation. Nodules can be found on the sea floor throughout the world's oceans, with surface coverage up to 70% in parts of the North Pacific Ocean. They are not confined to deep-sea but may also form in near-shore and fresh water environments. While predominantly occurring at the water/sediment interface, buried nodules have been recovered. The abundance of nodules in the upper 3 m of sediments is only 35% of the surface abundance. Ferromanganese nodules are found in oxidising conditions where sediment accumulation rates are low. Two problems remain which are relatively poorly understood. Firstly, nodules apparently grow more slowly than the local sedimentation rate yet remain surface manifestations. Secondly, the source of the manganese is controversial. Continental runoff, submarine volcanic activity and post-deposition migration out of the sedimentary interstitial water have all been suggested.

Considerable variation in the morphology of nodules exists. They may be spherical or discoid with diameters up to 15 cm. Discoid nodules have a definite identifiable orientation with respect to the sediment/water interface, with the convex surface exposed to sea water. The nodules form around a nucleation centre, and unusual examples often cited are shark's teeth and whale's earbones. Other morphological varieties include micronodules (a few millimetres in size), slabs, coatings and impregnations in rocks. The mineralogy of ferromanganese nodules consists primarily of hydrous manganese and iron oxides. While the nodules generally consist of amorphous (i.e. poorly crystalline) material, the principal manganese minerals which have been identified are birnessite [$4MnO_2 \cdot Mn(OH)_2 \cdot 2H_2O$], todorokite [$R^{2+}Mn_3O_7 \cdot 1\text{-}2H_2O$, where R^{2+} is a mixture of divalent cations], and nsutite or $\gamma\text{-}MnO_2$ [$(Mn^{2+}, Mn^{3+}, Mn^{4+})(O, OH)_2$]. In cross-section, the nodules have concentric light and dark bands related to iron and manganese concentrations, respectively. Nodules exhibit considerable enrichment of trace metals (see Table 5.8), and indeed it is this feature which is responsible for their mining potential. Mineralogical differences control the association of trace elements, presumably by cation substitution in the crystal structure. Manganese phases are enriched with copper, nickel, molybdenum and zinc, while iron phases typically exhibit enhancement of cobalt, lead, tin, titanium and vanadium. Gross chemical composition variations are observed. The manganese:iron ratio is greater than unity for nodules in the Pacific and Indian Oceans, but less than unity for Atlantic nodules. Near-shore marine nodules differ substantially from deep-sea varieties. The metals tend to occur in lower oxidation states, and the more rapid growth rates result in a higher organic content and lower concentrations of trace elements. This last feature makes near-shore nodules an unattractive mining proposition.

The formation of ferromanganese nodules in the deep-sea is a complex series of processes involving adsorption, oxidation and precipitation. Suitable nucleation sites are first coated by a layer of ferric oxide. This surface then acts as an adsorbent for Mn^{2+} which undergoes oxidative precipitation:

$$Mn^{2+} + 2OH^- + \tfrac{1}{2}O_2 \rightarrow MnO_2 + H_2O$$

Table 5.8. *Comparison of the average percentage composition of selected elements in crustal material with that in marine ferromanganese oxides*

Element	Crustal abundance (%)	FeMn oxide concentration (%)
Mn	0.095	16.174
Fe	5.63	15.608
Si	28.15	8.624
Al	8.23	3.098
Ca	4.15	2.535
Na	2.36	1.941
Mg	2.33	1.823
Ni	0.0075	0.489
Co	0.0025	0.299
Cu	0.0055	0.256
Mo	0.00015	0.041

Adapted from Cronan (1976).

Manganese dioxide is an efficient scavenger of iron and manganese, together with several trace metals. Adsorption from sea water thus acts as the mechanism by which trace element enrichment occurs. Nodules can continue to accumulate provided that an adequate supply of iron and manganese is maintained.

Halmyrolysates

Halmyrolysates are the alteration products resulting from reactions between sedimentary components and sea water. For lithogenous material encountering sea water, halmyrolysis is an extension of the chemical weathering processes initiated in the fresh water environment, with the major difference being the greatly enhanced concentrations of cations available for surface reactions. Such reactions may continue at the sediment/water interface prior to diagenesis while the sedimentary components are in contact with sea water. Consequently considerable overlap exists between the terms weathering/halmyrolysis and halmyrolysis/diagenesis.

During the transition from a fresh water to a sea water regime, the lithogenous material undergoes ion exchange leading to the precipitation of hydrogenous phases on the surfaces of the particles. Such processes are particularly important for the aluminosilicate components (i.e. clay minerals) which generally constitute the bulk of the riverborne continental material. As discussed in Section 2.6, clay minerals often exhibit a surface negative charge, and hence non-selective cation adsorption proceeds as part of the formation of an electric double layer. In this way, riverborne clay minerals entering sea water will take up Na^+ and Mg^{2+} in exchange for K^+ and Ca^{2+}. Alternatively, selective adsorption can occur and is manifest in the preferential uptake of cations with low hydration energies. Such cations (i.e. K^+, NH_4^+, Cs^+) may become fixed in inter-layer sites following dehydration and layer collapse. Reverse weathering reactions are examples of such mineralogical transformations. Montmorillonite may be transformed into illite following K^+ uptake. Sorbed cations may also become fixed in clay minerals by cation migration from the surface into the skeletal framework of the clay mineral, eventually occupying octahedral sites. Analogous anion migration of boron may cause its incorporation into the clay mineral. Marine sediments contain more boron than fresh water ones and therefore boron may be used as a tracer for saline environments in the geologic record.

5.5.3 Biogenous sediments

The biogenous components in marine sediments result from the fixation of mineral phases by marine organisms. With the exception of organic carbon, such mineral phases are fixed in the 'hard' parts of marine organisms. This means predominantly skeletal tests (i.e. shells) but teeth and scales may be significant for some components. Calcium carbonate and opaline silica are the most important biogenous constituents. The latter predominate beneath the most biologically productive surface waters (i.e. upwelling zones). Barite $[BaSO_4]$ and apatite $[Ca_5(F, Cl)(PO_4)_3]$ deposition may be mediated by biological processes. The former may be incorporated into tests while the latter occurs in fish teeth and scales. The relative importance of any biogenous component in the sediments depends upon the rate of production in surface waters and the subsequent preservation in both the water column and within the sediments. Clearly production rates are controlled by biological processes, while the preservation of biogenous material is determined by chemical processes. Factors influencing both the production and preservation of biogenous components in the sediments will be briefly outlined.

Calcium carbonate

Calcium carbonate of biogenous origin is a major component of marine sediments with almost one half of the surface sediments containing >30% $CaCO_3$. The carbonate is derived largely from the skeletal remains of foraminifera, coccolithophores and pteropods. Generally the greatest flux of material comes from organisms inhabiting the surface waters as opposed to deep-water and *benthic* (i.e. bottom-dwelling) species.

Pteropods are *pelagic* (i.e. inhabit the water column) molluscs which secrete tests of aragonite. They contribute significant quantities of calcium carbonate only to sediments in the Atlantic Ocean. Pteropod oozes cover approximately 2.4% of the Atlantic Ocean floor.

Coccolithophores are calcareous plantonic algae. The test is a *coccosphere* composed of a number of small calcareous plates called *coccoliths*. Upon death of the organism, the coccosphere disintegrates and the coccoliths are deposited in the sediments. These then comprise calcite particles in the size range 1 to 50 μm.

Foraminifera contribute the greatest quantity of calcium carbonate to the sediments. These are protozoans which secrete chambered tests about 0.5 to 1 mm in size. The morphology, especially the coiling orientation of successive chambers, and porosity of the tests yield information concerning water temperatures during growth and hence may be utilised as palaeoclimatic indicators in the geologic record. The most abundant species in the sediments are pelagic organisms which secrete calcitic tests. Benthic foraminifera may produce calcitic, arago-nitic or agglutinated (i.e. small grains of lithogenous material cemented with calcium carbonate) tests.

Upon the death of an organism, the test falls through the water column. Initially the calcareous test may be protected from dissolution due to the presence of organic envelopes or coatings of aluminium or iron oxides. Eventually the test is exposed to sea water and dissolution is initiated. There is a relatively well-defined depth in the ocean at which appreciable dissolution effects are noticeably for assemblages of tests known as the *lysocline*. The depth of the lysocline is related to water chemistry and is independent of the flux of calcareous material from the surface waters. Furthermore, the lysocline may be specified for each assemblage of organism giving rise to a foraminiferal lysocline and coccolith lysocline.

The solubility of calcium carbonate in sea water is governed by the equilibrium:

$$CaCO_3(s) \rightleftharpoons Ca^{2+} + CO_3^{2-}$$

Assuming, as discussed in Section 2.4, the activity of a pure solid to be unity, the theoretical ion activity product (IAP) is given by the solubility product:

$$K_{SP} = a_{Ca^{2+}} \cdot a_{CO_3^{2-}}$$

Comparison of the K_{SP} with the measured IAP of sea water defines three possible circumstances.

(i) $IAP/K_{SP} = 1$, system at equilibrium;
(ii) $IAP/K_{SP} > 1$, calcium carbonate supersaturated, precipitation anticipated;
(iii) $IAP/K_{SP} < 1$, calcium carbonate undersaturated, dissolution should occur.

While surface waters tend to be near equilibrium or supersaturated with respect to calcium carbonate, deep-waters are undersaturated. Factors which affect K_{SP} will be outlined.

The solubility product increases with an increase in pressure. The K_{SP} at 400 atmospheres (i.e. 4 km depth) equals $1.65 \times K_{SP}$ at 1 atmosphere. Similarly, the K_{SP} at 1000 atmospheres (10 km depth) equals $2.69 \times K_{SP}$ at 1 atmosphere. The K_{SP} is also temperature dependent, and, in the case of $CaCO_3$, unusually increases with a decrease in temperature. The total concentration of inorganic carbon can also exert an influence. Consider the following equilibrium:

$$CaCO_3(s) + CO_2 + H_2O \rightleftharpoons Ca^{2+} + 2HCO_3^-$$

Photosynthesis (confined to surface waters) utilises CO_2, and by Le Chatelier's principle will promote the reverse reaction causing precipitation. Alternatively, respiration produces CO_2 and dissolution of $CaCO_3$ will result. In summary, several factors lead to under-saturation of calcium carbonate at depth, namely a drop in temperature or an increase in pressure, which affect K_{SP}, and an increase in the partial pressure of carbon dioxide.

The crystalline form also affects the solubility of calcium carbonate. Marine organisms secrete calcite and aragonite, both of which are observed in marine sediments. Calcite is more thermodynamically stable and therefore has a lower solubility product:

$$K_{SP}(\text{calcite}) < K_{SP}(\text{aragonite})$$

While calcite should precipitate from sea water before aragonite, super-saturation of calcite may occur ultimately leading to aragonite precipitation. Three zones may be identified in the ocean: a surface layer in which aragonite is saturated, an intermediate layer in which calcite but not aragonite saturation occurs, and a deep layer in which calcite is under-saturated. The *calcium carbonate compensation depth* delineates the depth at the sediment/water interface for which the rate of $CaCO_3$ dissolution exceeds the rate of supply. Sediments accumulating below this depth will contain $<10\%$ $CaCO_3$.

Silica

Opaline silica ($SiO_2 \cdot nH_2O$) is secreted by several marine organisms including diatoms, radiolaria, silicoflagellates and sponges. Radiolaria and diatoms are the most important organisms and predominate in the sediments beneath the fertile waters associated with the equatorial Pacific and sub-Polar regions, respectively. Diatoms are phytoplankton with tests in the size range 10–100 μm. Radiolaria are zooplankton with tests 50–400 μm in size. As with calcareous tests, silicious tests may be protected from dissolution by organic envelopes or metal oxide coatings; however, dissolution is initiated after death. The water depth has no apparent effect on dissolution equilibrium and there is no compensation depth analogous to that of calcareous tests.

5.5.4 Diagenesis

The term diagenesis refers to the processes responsible for the alteration of a sediment following deposition. We shall ignore phenomena such as the recrystallisation of clay minerals and the crushing of large grain material, resulting from the extreme pressure associated with deep burial. From an environmental point of view, processes occurring within the upper few metres of the sedimentary column are the most important as they may exert influences upon the composition of the overlying water. At the time of deposition, sediments incorporate circumambient fluid to form *interstitial waters* (i.e. the aqueous solution occupying the cavities between sedimentary particles). Of particular interest in the diagenesis of sediments are the temporal changes in the composition of the interstitial waters resulting from bacterial activity, sediment/interstitial water processes or interstitial water/sea water interactions.

Diagenetic processes may be physical, chemical or biological in nature. An important physical process is compaction, whereby water is expelled from the sediments due to deposition of overlying sediment. This is manifest by a decrease in the porosity, Φ, of the sediment with depth. The porosity or volume fraction of interstitial water is given as:

$$\Phi = \frac{V_{IW}}{V_{IW} + V_S}$$

where

Φ = porosity,
V_{IW} = volume of interstitial water,
V_S = volume of solids.

Chemical processes include cementation, mineral segregation (i.e. layer formation) and ion-exchange processes. Cementation involves the chemical precipitation of a mineral phase which fills the cavities between sedimentary particles. Mineral segregation leading ultimately to layer formation results from the attainment of different equilibria at different depths with consequently strong chemical gradients. Bacterial mediation allows different equilibria to be established at different depths in the sedimentary column. In particular, biological processes may control the pH and pE within the interstitial waters. As discussed in

Section 2.5, these can be considered to be master variables that influence several equilibria. The following examples will serve to illustrate the importance of bacteria in the sediments and the chemical influences manifest in layer formations resulting from bacterial modifications of pH and pE.

Organic material is incorporated with other sedimentary components during deposition. In fresh water and coastal marine environments, the high biological productivity in the waters coupled to the rapid accumulation of sediments ensures a relatively high flux of organic material to the sediments. This acts as a major food source for benthic organisms, including bacteria. Biochemical oxidation of the organic material depletes the dissolved oxygen concentration of the interstitial waters, resulting eventually in the establishment of anoxic conditions within the sediments. The oxic/anoxic boundary occurs where the downward diffusion of oxygen balances the respiratory consumption. In a fashion analogous to that outlined for anoxic conditions in sea water (Section 5.3), the stepwise depletion of nitrate, nitrite and sulphate results in the accumulation of sulphide in the interstitial waters of marine sediments. This contrasts with fresh water environments which have little sulphate initially present. Hence, organic matter is used as an oxidant for respiration, and methane concentrations in anoxic interstitial waters increase.

Bacterial activity in near-shore sediments creates reducing conditions underlying a surface oxic layer. Clearly equilibrium conditions differ drastically and may cause mineral segregation. This is exemplified by the distribution of manganese within the sediments (Figure 5.20). Consider the case where MnO_2 near equilibrium with well-oxygenated sea water is deposited. Subsequent burial ensures disequilibrium is unavoidable. In the anoxic zone, the reductive dissolution of the solid mineral phase causes the dissolved manganese (Mn^{2+}) concentrations to increase. Enrichment over sea water content may be several orders of magnitude. A strong concentration gradient is established and manganous ions diffuse upward. Upon encountering oxic conditions, the manganous ions may be oxidatively precipitated from the interstitial waters.

The steady state profile for solid phase manganese indicates an exponential decrease in concentration with depth. Maximum surface concentrations arise due to the oxidative precipitation of manganese. The constant value at depth is indicative of $MnCO_3$ precipitation. The dissolved manganese profile in the interstitial waters exhibits a sub-

Figure 5.20. Schematic profiles of manganese dissolved in the interstitial waters (broken line) and associated with the solid sedimentary components (solid line).

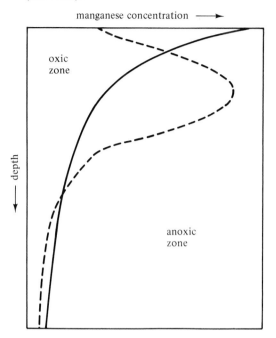

surface maximum. This occurs where the rate of supply of Mn^{2+} from the reductive dissolution of Mn(IV) oxides balances the rate of Mn^{2+} removal by upward diffusion and subsequent oxidative precipitation. Dissolved manganese concentrations decrease below this horizon due to $MnCO_3$ precipitation and the diminishing supply of Mn(IV) oxides in the sediments.

Finally, it should be noted that total lead profiles in sediments parallel those outlined here for manganese. However, these are ascribed to the increased anthropogenic input since the industrial revolution. While this is undoubtedly the case, caution must be exercised when investigating other elements. Tin profiles also exhibit an exponential decrease with depth. The comparison with lead and the known industrial importance of tin has led some workers to suggest a strictly anthropogenic origin, thereby ignoring potentially important diagenetic effects.

Questions

(1) What are the major products of a photochemical smog? Show chemical mechanisms by which these are formed.

(2) Explain how strong acids can be formed in the atmosphere.

(3) Describe simple chemical mechanisms accounting for the depletion of stratospheric ozone by supersonic aircraft emissions, use of chlorofluorocarbons and soil-derived dinitrogen oxide.

(4) Distinguish between congruent and incongruent weathering mechanisms for rocks. Give some examples of specific chemical mechanisms.

(5) What are the major ions in sea water, and what are their residence times? Which ions are responsible for buffering the pH and by what mechanism?

(6) Write an essay upon the chemical characteristics of soils.

(7) Discuss the chemical mechanisms involved in soil acidification. Why is aluminium mobilisation considered harmful to the environment?

(8) Define the terms lithogenous, hydrogenous, biogenous and chemogenous. What is meant by diagenesis?

(9) Describe factors influencing the dissolution and precipitation of calcium carbonate in the oceans at different depths.

(10) Define residence time, and express typical residence times for environmental chemical species. Report the longest and shortest residence times that you can identify and explain the factors which control them.

References

Bilinski, H. & Stumm, W. (1973). *EAWAG News* **2**, 3.

Brewer, P. (1975). 'Minor elements in sea water', in *Chemical Oceanography* (eds. J.P. Riley & G. Skirrow), 2nd edn., vol. 1, pp. 415–96. London: Academic Press.

Bruland, K. (1980). *Earth Planet Sci. Lett.* **47**, 176–98.

Cronan, D.S. (1976) 'Manganese nodules and other ferro-manganese oxide deposits', in *Chemical Oceanography* (eds. J.P. Riley & R. Chester), 2nd edn., vol. 5. London: Academic Press.

Dyrssen, D. & Wedborg, M. (1974). 'Equilibrium calculations of the speciation of elements in seawater', in *The Sea* (ed. E. Goldberg), vol. 5, pp. 181–95. New York: John Wiley & Sons.

Garrels, R.M. & Mackenzie, F.T. (1971). *Evolution of Sedimentary Rocks*. New York: W.W. Norton.

Gibbs, R.J. (1970). *Science* **170**, 1088–90.

Gregory, R. & Jackson, P. (1983). *Water Research Centre Regional Seminar on Reducing Lead in Drinking Water*.

Martin, J.H. & Knauer, G.A. (1980). *Earth Planet Sci. Lett.* **51**, 266–74.

Martin, J.M. & Meybeck, M. (1979). *Mar. Chem.* **7**, 177–206.

Salomons, W. and Forstner, U. (1980). *Environ. Technol. Lett.* **1**, 506–17.

Further reading

Brimblecombe, P. (1986). *Air*: *Composition and Chemistry*. Cambridge University Press.

Cresser, M. & Edwards, A. (1987). *Acidification of Freshwaters*. Cambridge University Press.

de Mora, S.J. & Harrison, R.M. (1984). *Chem. Brit.* **20**, 900–6.

Drever, J. (1988). *The Geochemistry of Natural Waters*, 2nd edn. Englewood Cliffs, NJ: Prentice Hall.

Harrison, R.M. (1990). *Pollution*: *Causes*, *Effects and Control*, 2nd edn. London: Royal Society of Chemistry.

O'Neill, P. (1985). *Environmental Chemistry*. London: George, Allen and Unwin.

Raiswell, R.W., Brimblecombe, P., Dent, D.L. & Liss, P.S. (1980). *Environmental Chemistry*. London: Edward Arnold.

APPENDIX

○ ○ ○ ○ ○ ○ ○ ○ ○ ○ ○ ○ ○ ○ ○ ○ ○ ○ ○ ○

Relative atomic masses of the elements

1	hydrogen	H	1.00797		35	bromine	Br	79.904
2	helium	He	4.0026		36	krypton	Kr	83.80
3	lithium	Li	6.939		37	rubidium	Rb	85.47
4	beryllium	Be	9.0122		38	strontium	Sr	87.62
5	boron	B	10.811		39	yttrium	Y	88.905
6	carbon	C	12.01115		40	zirconium	Zr	91.22
7	nitrogen	N	14.0067		41	niobium	Nb	92.906
8	oxygen	O	15.9994		42	molybdenum	Mo	95.94
9	fluorine	F	18.9984		43	technetium	Tc	
10	neon	Ne	20.179		44	ruthenium	Ru	101.07
11	sodium	Na	22.9898		45	rhodium	Rh	102.905
12	magnesium	Mg	24.305		46	palladium	Pd	106.4
13	aluminium	Al	26.9815		47	silver	Ag	107.868
14	silicon	Si	28.086		48	cadmium	Cd	112.40
15	phosphorus	P	30.9738		49	indium	In	114.82
16	sulphur	S	32.064		50	tin	Sn	118.69
17	chlorine	Cl	35.453		51	antimony	Sb	121.75
18	argon	Ar	39.948		52	tellurium	Te	127.60
19	potassium	K	39.102		53	iodine	I	126.9044
20	calcium	Ca	40.08		54	xenon	Xe	131.30
21	scandium	Sc	44.956		55	caesium	Cs	132.905
22	titanium	Ti	47.90		56	barium	Ba	137.34
23	vanadium	V	50.942		57	lanthanum	La	138.91
24	chromium	Cr	51.996		58	cerium	Ce	140.12
25	manganese	Mn	54.9380		59	praseodymium	Pr	140.907
26	iron	Fe	55.847		60	neodymium	Nd	144.24
27	cobalt	Co	58.9332		61	promethium	Pm	
28	nickel	Ni	58.71		62	samarium	Sm	150.35
29	copper	Cu	63.546		63	europium	Eu	151.96
30	zinc	Zn	65.37		64	gadolinium	Gd	157.25
31	gallium	Ga	69.72		65	terbium	Tb	158.924
32	germanium	Ge	72.59		66	dysprosium	Dy	162.50
33	arsenic	As	74.9216		67	holmium	Ho	164.930
34	selenium	Se	78.96		68	erbium	Er	167.26

69	thulium	Tm	168.934		81	thallium	Tl	204.37
70	ytterbium	Yb	173.04		82	lead	Pb	207.19
71	lutetium	Lu	174.97		83	bismuth	Bi	208.980
72	hafnium	Hf	178.49		84	polonium	Po	
73	tantalum	Ta	180.948		85	astatine	At	
74	tungsten	W	183.85		86	radon	Rn	
75	rhenium	Re	186.2		87	francium	Fr	
76	osmium	Os	190.2		88	radium	Ra	
77	iridium	Ir	192.2		89	actinium	Ac	
78	platinum	Pt	195.09		90	thorium	Th	232.038
79	gold	Au	196.967		91	protactinium	Pa	
80	mercury	Hg	200.59		92	uranium	U	238.03

INDEX